Feed Additives - Recent Trends in Animal Nutrition

Edited by László Babinszky

Published in London, United Kingdom

Feed Additives - Recent Trends in Animal Nutrition
http://dx.doi.org/10.5772/intechopen.111000
Edited by László Babinszky

Contributors
Ahmed El-Baz, Ali M. Humam, Bin Feng, Carlene Nagy, Ferenc Gál, Georgina Pesti-Asbóth, Ho Trung Thong, Ho Viet Duc, Ildikó Noémi Kovács-Forgács, Jen-Chieh Yang, Jiangfeng He, Judit Remenyik, Katerina Theodoridou, Le Nu Anh Thu, Lerato Deirdre Sehlabela, László Babinszky, Majdi A. Kairalla, Mammikele Tsatsimpe, Maria E. Rodriguez- Espinosa, Maria Eugenia Rodriguez Espinosa, Masindi Lottus Mphaphathi, Mohamed I. Idris Alshelmani, Nelson E. Ward, Nguyen Vu Thuy Hong Loan, Orsolya Csötönyi, Peiqiang Yu, Pulane Sebothom, Quanhui Peng, Raafat Khidr, Salah A. El-Safty, Samadi, Tao Ran, Taufiq Hidayat, Thivhilaheli Richard Netshirovha, Veronika Halas, Victor Guevara, Weixian Zhang, Xiaogang Yan, Zwivhuya Constance Rphalalani

First published in London, United Kingdom, 2024 by IntechOpen
IntechOpen is the global imprint of INTECHOPEN LIMITED, registered in England and Wales, registration number: 11086078, 167-169 Great Portland Street, London, W1W 5PF, United Kingdom

British Library Cataloguing-in-Publication Data
A catalogue record for this book is available from the British Library

Additional hard and PDF copies can be obtained from orders@intechopen.com

Feed Additives - Recent Trends in Animal Nutrition
Edited by László Babinszky
p. cm.

This title is part of the Veterinary Medicine and Science Book Series, Volume 20
Topic: Animal Nutrition
Series Editor: Rita Payan Carreira
Topic Editor: Manuel Gonzalez Ronquillo

Print ISBN 978-1-83769-081-7
Online ISBN 978-1-83769-082-4
eBook (PDF) ISBN 978-1-83769-083-1
ISSN 2632-0517

For EU product safety concerns:
IN TECH d.o.o., Prolaz Marije Krucifikse Kozulić 3, 51000 Rijeka, Croatia,
info@intechopen.com or visit our website at intechopen.com.

IntechOpen Book Series

Veterinary Medicine and Science

Volume 20

Aims and Scope of the Series

Paralleling similar advances in the medical field, astounding advances occurred in Veterinary Medicine and Science in recent decades. These advances have helped foster better support for animal health, more humane animal production, and a better understanding of the physiology of endangered species to improve the assisted reproductive technologies or the pathogenesis of certain diseases, where animals can be used as models for human diseases (like cancer, degenerative diseases or fertility), and even as a guarantee of public health. Bridging Human, Animal, and Environmental health, the holistic and integrative "One Health" concept intimately associates the developments within those fields, projecting its advancements into practice. This book series aims to tackle various animal-related medicine and sciences fields, providing thematic volumes consisting of high-quality significant research directed to researchers and postgraduates. It aims to give us a glimpse into the new accomplishments in the Veterinary Medicine and Science field. By addressing hot topics in veterinary sciences, we aim to gather authoritative texts within each issue of this series, providing in-depth overviews and analysis for graduates, academics, and practitioners and foreseeing a deeper understanding of the subject. Forthcoming texts, written and edited by experienced researchers from both industry and academia, will also discuss scientific challenges faced today in Veterinary Medicine and Science. In brief, we hope that books in this series will provide accessible references for those interested or working in this field and encourage learning in a range of different topics.

Meet the Series Editor

Rita Payan Carreira earned her Veterinary Degree from the Faculty of Veterinary Medicine in Lisbon, Portugal, in 1985. She obtained her Ph.D. in Veterinary Sciences from the University of Trás-os-Montes e Alto Douro, Portugal. After almost 32 years of teaching at the University of Trás-os-Montes and Alto Douro, she recently moved to the University of Évora, Department of Veterinary Medicine, where she teaches in the field of Animal Reproduction and Clinics. Her primary research areas include the molecular markers of the endometrial cycle and the embryo–maternal interaction, including oxidative stress and the reproductive physiology and disorders of sexual development, besides the molecular determinants of male and female fertility. She often supervises students preparing their master's or doctoral theses. She is also a frequent referee for various journals.

Meet the Volume Editor

László Babinszky is a professor of animal nutrition. Currently, he is a professor emeritus at the University of Debrecen and at the Hungarian University of Agriculture and Life Sciences, Kaposvár Campus, Hungary. He earned his Ph.D. from Wageningen University in the Netherlands. He has worked in various universities and research institutes in Hungary, the Netherlands, Austria, and Germany. Dr. Babinszky has authored more than 300 publications (papers and book chapters), and edited four printed books, and authored and edited four e-books. He is a member of various international committees (e.g., he has been a member for 27 years of the American Society of Animal Science [ASAS], USA) and various international journal editorial boards (e.g., *Acta Agriculturae Scandinavica, Section A—Animal Science*).

Contents

Preface

One of the biggest challenges of animal agriculture in the twenty-first century is how to supply the Earth's growing population with high-quality, healthy, and safe food of animal origin. Nowadays, it is also a legitimate requirement that the animal origin foodstuffs be produced in a sustainable manner. To solve this task, many questions need to be clarified, but one of the most important is how we can further improve the biological, technological, and economic efficiency of animal nutrition.

One possible way to improve biological efficiency is the professional use of feed additives, which include the latest biological, biochemical, and physiological knowledge and the use of state-of-the-art laboratory analytical methods. This book focuses on some newer aspects of the various feed additives (vitamins, enzymes, acidifiers, and various plant feed additives) in poultry, pig, and ruminant nutrition. A further aim is to demonstrate the new trends in vitamin nutrition and the relationship among the metabolic disease and production level and the application of plant feed additives in broiler nutrition. Furthermore, the book also shows the use of molecular spectroscopy when an enzyme is added to the ruminant diets and how to mitigate ruminant methane emission with phytochemicals (plant feed additives).

This book is recommended for scientists, graduate students, and those working in animal agriculture. We hope that readers will find useful information in this book for their daily work or studies. If this is so, then our efforts were not in vain.

I would like to thank all the chapter authors for their excellent contributions. I'm also grateful to the staff at IntechOpen, particularly Publishing Process Manager Karmen Daleta and Commissioning Editor Lucija Tomicic-Dromgool.

Dr. László Babinszky
Professor,
Faculty of Agricultural and Food Sciences and Environmental Management,
Department of Animal Nutrition and Physiology,
University of Debrecen,
Debrecen, Hungary

Department of Farm Animal Nutrition,
Institute of Physiology and Nutrition,
Hungarian University of Agriculture and Life Science,
Kaposvár Campus,
Kaposvár, Hungary

Chapter 1

Introductory Chapter: Requirements for Feed Additives in the Twenty-First Century

László Babinszky

1. Introduction

1.1 Definition and classification of feed additives

Feed additives have been used by farmers since ancient times, when, for example, salt was added to animal feed to improve its taste. In the beginning, farmers added natural products such as plants, minerals, and waste products to feed [1].

Later, as technology and knowledge of physiology and biochemistry advanced, farmers and nutritionists used synthetic vitamins, minerals, and growth promoters as feed additives in animal husbandry.

Feed additives are defined in the literature in several ways. Due to limited space, let me present here only two examples:

According to the European Commission [2]: "Feed additives are products used in animal nutrition for purposes of improving the quality of feed and the quality of food from animal origin, or to improve the animals' performance and health, e.g. providing enhanced digestibility of the feed materials. Feed additives may not be put on the market unless authorization has been given following a scientific evaluation demonstrating that the additive has no harmful effects, on human and animal health and on the environment."

As defined by EFSA [3]: "Feed additives are products used in animal nutrition to achieve an effect on the feed itself, on the animals, on food products obtained from the animals consuming the feed additive, or on the environment. For instance, feed additives are used to enhance flavor of feed, to meet the need for certain nutrients, or to increase the performance of animals in good health. They are used in feed for food-producing animals and in pet food."

I think that these two definitions summarize well what the term feed additive means. Feed additives are products that improve the digestibility and palatability of feed as well as the performance and health status of animals and the quality of animal products. The most important function of some feed additives is to reduce the environmental impact of animal agriculture, in such a way that the production of livestock does not decrease.

In addition to these, it is an important requirement that these additives cannot be harmful to human consumers, animal health, and the environment.

IntechOpen

The use of feed additives also has an important economic condition, i.e., their use must not increase the price of feed to such an extent that the profitability of production is endangered.

Feed additives can be grouped in several ways. According to one of the most accepted classifications, there are the following feed additives [3, 4].

- Sensory additives (feed additives affecting the sensory properties of animal products e.g., flavors and colorants);
- Technological additives (preservatives, antioxidants, substances decreasing mycotoxin contamination of feeds, emulsifiers, acidity regulators, silage additives, etc.);
- Zootechnical additives (immunomodulators, digestive stimulants, growth promoters of non-microbial origin, substances increasing performance or quality of animal products, etc.);
- Nutritional additives (vitamins, amino acids, trace elements, minerals, plant enzymes, etc.).
- Coccidiostats and histomonostats.

2. Global market of feed additives

Statistical data and various forecasts show that the feed additive market is continuously growing and this trend will also be continued in the future. Forecasts show that significant growth is expected in the feed additives market in the coming years, mainly due to the increase in demand for pigs, sheep, poultry, and aquaculture [5].

According to FMI [6], the global animal feed additive market will grow from USD 55,842.2 million in 2024 to USD 109,184.5 million by 2034.

According to forecasts, the spread of plant-derived substances in particular will be significant in the coming year due to the ban on the use of certain antibiotics, harmful residual effects, and better cost-effectiveness [7].

This fact imposes additional tasks on feeding research and development.

3. Laboratory and animal experiments with feed additives according to today's expectations

In the past, farmers and nutritionists decided only on the basis of practical experience which additive to add to the feed and in which concentration. However, nowadays this is no longer an acceptable method.

Today, one of the biggest issues in animal nutrition is how to feed farm animals so that the animal product (e.g., meat, milk, eggs, etc.) is safe for the consumer and the environment is not further burdened during animal production. One of the most important questions is in this regard, how to use feed additives in animal nutrition.

In order to develop a safe and environmentally friendly additive, very comprehensive laboratory analysis and animal experiments must be carried out.

It is also very important to know what is/are the active substance/substances in the feed additive. To this end, a reliable and accurate analytical method for measuring the active substance must be available.

In order to develop a safe feed additive, the metabolism (pathway) of the active substance must also be known. Furthermore, it must be ensured that no harmful intermediates are produced in the animal's body during the metabolism of the active substance.

In all cases, the applicability and effectiveness of the given feed additives must be verified by animal studies in accordance with international standards. The repeatability of the treatment effect must also be confirmed by the results of animal experiments and their statistical analysis.

It is recommended to verify the results of animal studies obtained at the research site/development site and also in the practice (on the farm).

In addition to laboratory and animal experiments, it is very important that the feed additive can be homogeneously mixed into the compound feed. If the homogeneity value is adequate, we can be more or less certain that the daily feed additive intake of the animals is close to the planned value. It should be noted, that the homogeneity and absorption of the active substance can be favorably improved by the various nanotechnological processes available today.

The homogenization of the active substance in the compound feeds can be ensured by various technological procedures, including the application of nanotechnology.

A very important additional requirement is that, during mixing feed additive into compound feed, the active substance of the additive is not damaged during various physical treatments.

Summarizing the above, different feed additives can only be used in animal nutrition if:

- a suitable laboratory method is available to measure the concentration of the active substance in the feed additive and in the compound feed;

- the pathway of the active substance in the animal metabolism is known;

- the feed additive can be accurately and homogeneously mixed into the compound feed;

- the positive effect of the additives has been proven by animal experiments;

- the use of the feed additive is also justified from an economic point of view;

- the feed additive is demonstrably safe for both animals and the consumers.

4. The aim of the book

This book focuses on some newer aspects of the various feed additives (vitamins, enzymes, acidifiers, and various plant feed additives) in poultry, pig, and ruminant nutrition.

A further aim is to demonstrate the new trends in vitamin nutrition and the relationship among the metabolic disease and production level and application of plant feed additives in broiler nutrition.

Furthermore, the book also shows the use of molecular spectroscopy, when enzyme is added to the ruminant diets and how to mitigate ruminant methane emission with phytochemicals (plant feed additives).

This book is recommended for scientists, graduate students, and those working in the animal agriculture.

We do hope that readers will find useful information in this book for their daily work or studies.

If this is so, then our efforts were not in vain.

Acknowledgements

I would like to thank all the chapter authors for their excellent contributions. I'm also grateful to the staff at IntechOpen, particularly Publishing Process Manager Karmen Daleta and Commissioning Editor Lucija Tomicic-Dromgool.

Author details

László Babinszky[1,2]

1 Faculty of Agricultural and Food Sciences and Environmental Management, Department of Animal Nutrition Physiology, University of Debrecen, Debrecen, Hungary

2 Department of Farm Animal Nutrition, Institute of Physiology and Nutrition, Hungarian University of Agriculture and Life Science, Kaposvár, Hungary

*Address all correspondence to: babinszky@agr.unideb.hu

References

[1] Wani FF, Asimi OA, Khan IA. Use of feed additives in animal nutrition. Journal of Scientific Research and Reports. 2023;**9**:82-87

[2] European Commission. Feed Additives; 2023. Available from: https://food.ec.europa.eu/safety/animal-feed/feed-additives_en

[3] EFSA (European Food Safety Authority). 2023. Feed Additives. Available from: https://www.efsa.europa.eu/en/topics/topic/feed-additives

[4] Karásková K, Suchý P, Straková E. Current use of phytogenic feed additives in animal nutrition: A review. Czech Journal of Animal Science. 2015;**12**:521-530

[5] 360 RR (360 Research Report). 2023. Feed Additives Market Size in 2023: Growth Opportunities and Future Outlook 2030. Available from: Feed Additives Market Size In 2023: Growth Opportunities and Future Outlook 2030 (linkedin.com)

[6] FMI (Future Market Insights). 2023. Animal Feed Additive Market. Available from: Animal Feed Additives Market Trends & Forecast by 2034 | FMI (futuremarketinsights.com)

[7] Muneendra K, Vinod K, Debashis R, Roy K, Raju K, Shalini V. Application of herbal feed additives in animal nutrition—A review. International Journal of Livestock Research. 2014;**4**(9):1-8. DOI: 10.5455/ijlr.20141205105218

Chapter 2

Vitamin Supplementation in Broiler Feeds and U.S. Survey on Fortification Rates

Nelson E. Ward

Abstract

This chapter covers a short review of the vitamin discovery period, followed by a discussion of the vitamins as nutritional supplements for poultry diets. These organic molecules perform within a complex metabolic system, and function in catalytic, developmental, and protective roles. Research in recent years suggests vitamins also play a pivotal role in the intestinal microbiome and "gut health" and may have direct effects on the establishment of a more desirable microbial population. Rapid changes in poultry genetics requires modifications in fortification rates, especially when less feed is required to attain these improvements. A survey on the vitamin fortification rates of broiler feeds in the U.S. is also included for discussion and comparison with a similar 1993 survey and the National Research Council. Some vitamins showed a wider disparity in fortification levels than others.

Keywords: vitamin, broilers, laying hens, turkeys, feed additives

1. Introduction

Vitamins play a decisive role in the nutrition and health of all animals, including poultry. The requirements for vitamins and nutrients in general by poultry are affected variably by progressive genetics, disease, levels of production, and production systems. Production within an indoors environment places greater scrutiny on vitamin fortification levels in formulated feeds.

Many factors influence vitamin content of feed ingredients [1]. Whereas the level, stability, and bioavailability of vitamins in ingredients are highly variable and too low to meet requirements, commercial forms of vitamins offer greater stability, efficacy, and accuracy in dietary supplementation rates. Bioavailability and mixing characteristics can be maximized in synthetic forms that can be added to feeds in very small amounts (micrograms per tonne feed in some cases). Today, 12 individual vitamins are typically added to poultry feeds which provides nutritionists the opportunity to supplement each according to target.

As research reveals new information on the benefits of vitamin supplementation, this can impact vitamin fortification objectives. Studies on the effect of niacin on body phosphorus accumulation [2], or on the effect of the fat-soluble vitamins on carcass and meat quality [3], rooster fertility [4], and disease resistance [5], for example, can influence fortification rates in commercial feeds.

In 2017–2018, an unprecedented shortage in global vitamin supplies forced many nutritionists to scrutinize fortification guidelines as a means to control spiking costs and scarce supplies, a practice that most likely has affected today's fortification rates.

2. Brief perspective of the golden age of vitamins

The history of vitamin research and discovery is certainly some of the more remarkable in science. The twentieth century highlights the period during which the vitamins were first officially recognized as being essential in nutrition [6, 7]. Leading up to the early 1900s, scientists were cognizant of the fact that a diet composed of carbohydrate, fat, protein, and salts was incapable of preventing certain disease-like maladies. In a speech known as the "vitamin theory", Hopkins in 1906 noted that something in "astonishingly small amounts" was needed in animal diets beyond these basic dietary components [8].

Later, in 1912, Casimir Funk coined these yet-to-be-identified constituents as "vitamines" for "vital amines", a term changed to "vitamins" in recognition that these were not all amines [8]. Names such as Eijkman, Funk, Stepp, and Hopkins—and many more—are popular references to the early investigations and observations that laid the foundation in "vitamin science" of the 1930s and 1940s—a period considered as the 'Golden Age of Vitamins' [7, 8]. The discovery of 13 vitamin groups covers only a few decades, starting with the first vitamin in 1913 ("factor A" that was later renamed vitamin A) and culminating in 1948 when vitamin B_{12} was isolated and defined.

Yet, in his exhaustive review, McDowell [7] points out that some of the vitamin related maladies were recorded in Chinese literature as far back as 2600 B.C. That components of some foods and plants could cure many of these illnesses—broth of pine needles or juice of citrus fruits to cure scurvy, for example—was subtle acknowledgment that a basic diet lacked some nutritional but essential aspect. "Diseases" such as blindness, beriberi, scurvy, pellagra, and rickets went unrecognized as nutritional deficiencies by the earliest chemists, physiologists, and researchers. By 1900, only two or three of these were officially recognized to be associated with the diet.

In all, 17 vitamin-related Nobel Prizes were awarded from 1928 to 1967 on the discovery, isolation, synthesis, and structure of the vitamins [9]. In most cases, once the structure of the vitamin was chemically elucidated, the first synthesis was accomplished soon afterwards [10].

Today, economically important vitamin production occurs by chemical means, fermentation, or through the extraction from natural sources. Over the years, significant improvements in the stabilization and commercial product formulation were established in human and animal vitamin product forms. Advancements in the industrial production of vitamins makes possible the supplementation of vitamins of commercial animal diets in agriculture [1, 11]. The market size of the global vitamin supplement business was valued at USD 44.12 billion in 2020 and is expected to expand at a compound annual growth rate of 6.2% from 2021 to 2028 (https://www.grandviewresearch.com/industry-analysis/vitamin-supplements-market-report).

3. The vitamins

Vitamins comprise a group of organic compounds distinct from fats, carbohydrates, and proteins. They are considered to be (1) organic, (2) natural components of

Vitamin	Solubility	Primary functions
Vitamin A	Fat	Vision, reproduction, membranes, bone development, hatchability, ataxia and weakness, ruffled feathers
Vitamin D_3	Fat	Bone development (P, Ca absorption), immune function
Vitamin E	Fat	Antioxidant, cell membrane integrity, immune function, reduced platelet aggregation (blood clotting)
Vitamin K	Fat	Blood clotting, bone mineralization
Thiamin	Water	Energy production, and carbohydrate, fat and protein metabolism, nerve function
Riboflavin	Water	Energy production, and carbohydrate, fat and protein metabolism
Niacin	Water	Energy production, and carbohydrate, fat and protein metabolism
Pyridoxine	Water	Energy production, and carbohydrate, fat and protein metabolism
Pantothenate	Water	Energy metabolism
Biotin	Water	Carbohydrate, fat and protein metabolism, glucose metabolism
Folic acid	Water	Amino acid & energy metabolism, protein synthesis, immunity
Vitamin B_{12}	Water	Related to methionine, choline and folacin metabolism, and fat and carbohydrate metabolism

Table 1.
Twelve vitamins commonly added to poultry feeds and their primary functions.

foods in minute amounts, (3) essential for normal physiological function, (4) associated with distinct deficiency symptoms when absent, and (5) insufficiently produced by the host [6]. Today, we recognize 13 different vitamins that meet these standards, of which 12 are routinely supplemented to commercial poultry feeds.

These molecules perform within a complex metabolic system, and function in catalytic, developmental, and protective roles (**Table 1**). As such, vitamins are essential for growth, development, maintenance, and reproduction, and mediate in synthetic and degradative processes and participate in catalytic functions. As opposed to macro ingredients in animal feeds, vitamins are required in comparatively small amounts to satisfy requirements, thus are considered micro ingredients in the realm of commercial feed additives. Depending on the animal species, some vitamins are produced in the body (for example, niacin from tryptophan; choline from methionine; vitamin D_3 from 7-dehydrochlesterol via ultraviolet light; ascorbic acid by most animals, including poultry), albeit at levels insufficient to meet demand for metabolic purposes. Survival depends on the presence of vitamins.

4. Vitamins are not a chemical class

Unlike other chemical classes such as the alcohols or aldehydes, the term 'vitamin' does not refer to a class of chemicals with similar structures or functions. Individual vitamins vary significantly in chemical structure (**Figure 1**) but are categorized in two groups based on solubility: fat-soluble and water-soluble. The four fat-soluble vitamins include vitamins A, D_3, E and K, while the nine water-soluble vitamins comprise thiamin, niacin, riboflavin, pyridoxine, pantothenic acid, vitamin B_{12}, folic acid, biotin, and ascorbic acid (vitamin C). Generally, the fat-soluble vitamins are more

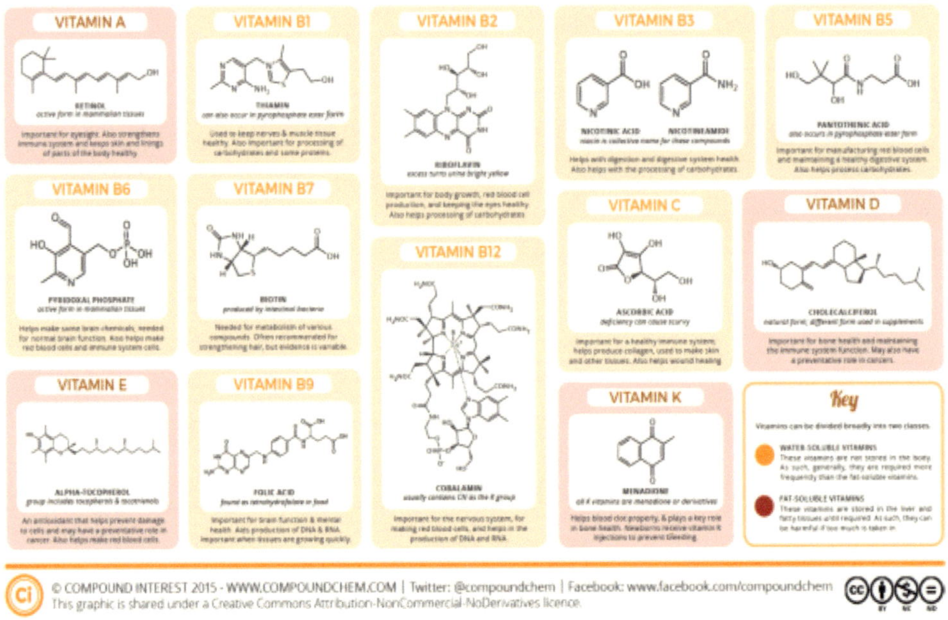

Figure 1.
The structure of vitamins varies considerably (www.compoundchem.com).

aligned with cell membrane components, whereas the water-soluble vitamins often function as carriers of several biochemical groups or act as coenzymes in metabolic reactions [12], albeit some exceptions do occur. Each vitamin is distinct in function and elicits different deficiency symptoms when absent or insufficient.

Fat-soluble Vitamins. These vitamins are primarily aliphatic and aromatic in nature and are absorbed in a manner similar to fats and oils. Being hydrophobic, they require emulsification in the upper intestinal tract of poultry through gizzard action and intestinal churning. Mixed micelles—a combination of free fatty acids, monoglycerides, and bile acids—deliver the fat-soluble vitamins to the microvilli surface for uptake into the portal circulation of poultry. For some animal species, regional intestinal differences generally exist for absorption: proximal for vitamin A, medial for vitamin D_3, and distal for vitamins E and K [13]. For poultry, the form of the vitamin can affect absorption site. Combs and McClung [6] note that vitamin D_3 absorption rate is fastest proximally, but owing to a longer feed transit time, the greatest amount probably occurs distally. Competition for absorption exists among the fat-soluble vitamins [13], which probably extends to the rate of accumulation for fat-soluble vitamins in eggs [14]. The magnitude to which interactions occur among the fat-soluble vitamins is impacted by supplementation levels in the feed [15].

Water-soluble Vitamins. Absorption is influenced by molecular weight, ionization status, and whether the vitamin is present as a weak acid or base [6, 12]. Intestinal uptake is favored by a small molecular structure and a weak ionic character. While niacin (niacinamide), pyridoxine, biotin and vitamin C are readily taken up, absorption of thiamin and B_{12} face greater difficulty. Thiamin, B_{12}, folic, and vitamin C are absorbed by a carrier-mediated mechanism, but as levels in the feed increase, simple diffusion plays a greater role [12]. Less is known about potential competition among water-soluble vitamins. These vitamins are somewhat unique from their fat-soluble counterparts in that their many interactions with one another in metabolism can make it difficult to determine some of the individual vitamin requirements.

Once absorbed, the vitamins function in five basic roles in metabolism, as noted by Combs and McClung [6], and these include antioxidants, gene transcription, electron donors and acceptors, hormones, and coenzymes. Respective chemical reactivity and tissue distribution come into play for the completion of these roles.

Vitamins are generally considered nontoxic. Minimal body storage occurs among the water-soluble vitamins. Whereas fat-soluble vitamins are generally stored in the liver and other tissues, vitamin K appears to be rapidly metabolized and excreted. Factors such as age, duration of feeding, and form of the vitamin can affect tolerance to high levels. For example, hypervitaminosis is less likely with the vitamin D_2 as opposed to the more potent vitamin D_3 in poultry [16]. Dietary calcium and phosphorus can also influence the tolerance to vitamin D. Furthermore, laying hens may be more tolerant to high levels of vitamins as opposed to broilers. While 2800 IU vitamin D_3/kg fed for more than 60 days is listed as upper limit for chickens [16], a level of 102,200 IU D_3/kg triggered no adverse symptoms for laying hens over a 40-week period [17]. In part, the higher tolerance presumably is associated with the transfer of excess vitamin D_3 from the body to eggs, thus avoiding build-up in the body. Combs and McClung [6] categorize vitamins in four groups of toxic potential –

- Greatest potential—vitamin A, vitamin D_3
- Moderate potential—niacin/niacinamide
- Low potential—vitamin E, vitamin C, thiamin, riboflavin, pyridoxine
- Negligible potential—vitamin K (menadione), pantothenic acid, biotin, folic acid, vitamin B_{12}

5. Commercial vitamin forms

Moisture, heat, pelleting, oxidation, and reduction reactions can undermine vitamin survivability in feeds. These stresses are often associated with pelleted, expanded, or extruded feeds, as well as minerals in the feeds, or premixes with other additives that are hygroscopic in character. In addition, the level and bioavailability of vitamins that naturally occur in ingredients can vary considerably [1]. Agronomic factors influence vitamin content, as does plant maturity and harvest conditions, and the stability of naturally occurring vitamins seldom holds up to the rigors of feed processing. As such, many fail to meet standard requirements for poultry without supplementation from other sources.

Commercial product forms of vitamins are formulated to buffer these challenges that can adversely influence the amount the animal eventually consumes in the feed. Since vitamins vary widely in their chemical structure, the type of formulation and final product form for each can differ. In the end, the goal is to develop commercial vitamins to optimize handling, mixability, stability, and bioavailability. Some of the more common formulations include beadlet formation, spray drying, adsorption, crystalline powder, or coated powder in the final formulation.

Improvements usually encompass chemical or physical modifications (**Table 2**). *Chemical modification* of vitamin A, E and C, can stabilize the reactive hydroxyl

Vitamin	Formulation	Purpose
Vitamin A	Ester in a cross-linked beadlet	Stability, solubility
Vitamin D3	Spray dry (SD), beadlet	Stability, uniformity
Vitamin E	Acetate ester, SD or granular	Flow, reduced dustiness
Vitamin K/menadione	Crystalline powder	Flow, handling
Thiamin	Coarse granular	Stability
Riboflavin	SD granular	Flow, handling
Pyridoxine	Fine granular crystals	Stability, mixing
Vitamin B_{12}	Crystalline w/carrier	Distribution
Niacin	Crystalline	Flow, reduced dustiness
Niacinamide	Crystalline	Flow, reduced dustiness
Ca-pantothenate	SD	Flow, reduced dustiness
Biotin	SD	Distribution, handling
Folic acid	SD	Flow, stability, mixing
Vitamin C	P esterification, ethyl cellulose coat	Stability

Table 2.
Formulated changes in vitamins to improve function for commercial purposes.

groups through esterification. Antioxidants may also be included for added protection against reactions that are prone to occur in the presence of other factors such as moisture and some minerals. *Physical protection* is applied in various formulations to develop a barrier to protect against oxygen, moisture, or light. Differences exist across manufacturers and formulation technologies due to patents and proprietary techniques, hence, not all will equally protect against degradation.

The protection of the vitamin A molecule is a good example of chemical and physical technologies being utilized for one vitamin. Once chemically stabilized through esterification, further improvements are made by cross-linking with gelatin in a 'beadlet'. Fructose and glycerine enhance the process to ensure protection against moisture and heat during feed manufacture. The protein-based coating can be hydrolyzed by low pH and intestinal proteases, thus releasing the vitamin A for absorption once consumed by the animal.

In addition to beadlet formation, improvements can also be made through encapsulation, adsorption and spray drying. Some vitamins are inherently stable, such as niacin or biotin, and require minimal formulation changes beyond grinding and sifting to improve particle size and particle distribution while reducing dustiness in the final form. Liquid or emulsified vitamins have some application but are not a common product form. For example, the liquid application to pelleted feeds may experience fewer losses via the pelleting process, which may allow nutritionists to finetune fortification rates. In addition, some costs-savings may occur by avoiding the costs to develop beadlets or product forms for improved stability.

6. Guidelines: Vitamin fortification in commercial feeds

Commercial vitamin supplementation rates of poultry feeds are provided by several sources, ranging from formal guidelines and university research to standards developed in field trials in commercial production. Guidelines or recommendations for vitamins are usually in addition to levels already present naturally in the ingredients being fed. Vitamins that occur in ingredients vary widely in bioavailability, as well as stability during pelleting of feeds.

The National Research Council (NRC) Nutrient Requirements for Poultry is considered one of several formal standards for the requirement of nutrients. Twelve vitamins are listed at levels summarized from research studies from published results of nutritional research (**Table 3**) [18]. Besides being quite dated, the most recent NRC [18]—and similar guidelines from other sources—excludes consideration for margins of safety to account for various factors that could influence vitamin requirements. Considerations for other factors—vitamin form, inadequate mixing, poor storage conditions, malabsorption issues, genetic change, lower feed intake, and stress status—are not considered, yet these factors can impact the supplemental levels necessary to attain optimal performance and production in commercial practice.

To this end, Applegate and Angel [19] note that "our perception and definition of a nutrient requirement has changed from first being a requirement, as a percent of a diet, to preventing a nutrient deficiency, to now being a requirement to optimize growth or egg production response per unit of nutrient intake." Supplementation rates for nutrients as a feed additive can be adjusted to any particular goal: rapid growth (for example, the greatest body weight within a given time period) or optimal growth (where the conversion of feed to body weight gain becomes the primary goal), or to achieve the lowest cost per unit of feed or lowest cost per unit of meat produced. Economics play a dominant role in research objectives to define the requirement of vitamins in many cases.

The rapid change in broiler genetics is generally the most influential factor that affects long-term fortification levels. Broiler growth rates have improved 3–4% annually with less feed being consumed/unit live gain [20]. To this end, the feed conversion ratio (FCR) improved nearly 1.5% annually over the past 10 years [21]. Patricio et al. [22] presented calculations to show that 20% less feed was required to attain the

Units/MT feed*	Starter	Grower	Finisher
Vitamin A, MIU	1.5	1.5	1.5
Vitamin D_3, MIU	0.2	0.2	0.2
Vitamin E, TIU	10	10	10
Vitamin K, G	0.5	0.5	0.5
Niacin, G	27	27	11
Thiamin, G	1.8	1.8	1.8
Riboflavin, G	3.6	3.6	3.6
Pyridoxine, G	3	3	2.5
Pantothenic, G	10	10	10
Vitamin B_{12}, MG	9	9	9
Folic, MG	550	550	250
Biotin, MG	150	150	100

**MIU = million international units; TIU = thousand international units; G = grams; MG = milligrams.*

Table 3.
Vitamin recommendations [18] for broilers.

same body weight in broilers over the span of nearly 20 years. Alternatively, the time required to attain a 2.31 kg slaughter weight has declined from 52 days 1995 to 40 days today [21]. Modern-day caged layers are considered "long life" layers, based on the production of 500 eggs in 100 weeks [23]. If dietary vitamin levels were held steady, for example, vitamin intake declines per unit feed intake while the requirement is increased for vitamins and other nutrients because of the higher production. So, founded simply on improvements in the bird's ability to produce more with less feed, micro-nutrients such as vitamins require adjustments to offset a reduced feed intake.

Sources for vitamin fortification guidelines are offered by organizations other than the NRC. Genetic companies for poultry periodically conduct research to determine the most optimal nutrient levels for their genetic base, and this includes vitamins [24, 25]. Likewise, commercial vitamin suppliers present recommendations for vitamin supplementation rates [1, 26]. The Optimum Vitamin Nutrition®, for example, proactively considers commercial stresses and conditions that influence vitamin supplementation of poultry feeds [1, 26]. This concept targets the health and productivity of poultry over a range of vitamin supplementation levels, as noted in **Figure 2** (taken from [1, 26]). In this approach, average animal response refers to the animal response in terms of productivity: feed conversion, growth rate, etc., as a consequence of vitamin consumption. Total vitamin intake considers amount of vitamin provided in the diet from all sources and considers the bioavailability of that vitamin in feedstuffs. Deficient refers to a deficiency status relative to recommendations by the NRC and other similar sources. Sub-optimum vitamin intake results in subpar health and productivity and prevents clinical deficiencies, whereas OVN intended to compensate for factors that can negatively impact animal health, well-being, and productivity. While it is difficult to account for all factors that affect animal performance, this program considers many prominent factors that could influence vitamin supplementation.

There is no shortage of evidence that targeted vitamin fortification can be beneficial. Luo et al. [27] highlight the benefits of higher vitamin supplementation for broilers exposed to coccidiosis under commercial conditions, in part, because of lower digestibility by diseased birds. In addition, folic acid at a level roughly 10-fold higher

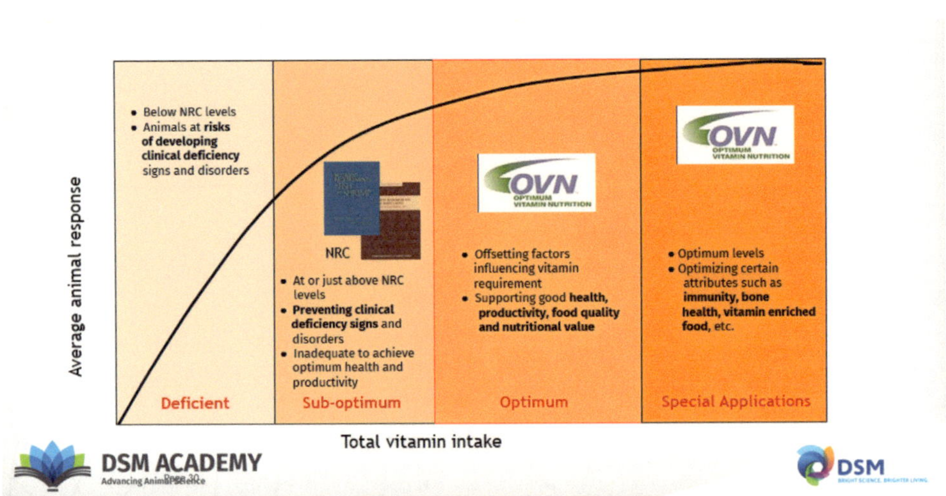

Figure 2.
Optimum vitamin nutrition concept over the range of deficiency to special applications.

than average fortification rates in the U.S. increased gene expression for more muscle accretion in broilers [28]. On the other hand, more breast meat (*P. major*) accumulation occurred with 25-OH D_3 *in ovo* injections [29], which is consistent with an improved mitotic satellite cell formation and breast muscle yield in broilers fed 25-OH D_3 [30]. Higher fortification is required to enhance egg levels of some vitamins for human consumption such as vitamin D_3 [17, 31] and folic acid [32]. So, in effect, vitamin fortification strategies or "requirements" are modified, depending on an improved return on investment, or to take advantage of effects in meat or eggs, or other endpoints [33].

7. Vitamins and intestinal microflora

Studies have illuminated an important relationship between vitamin fortification of feeds and intestinal microflora in mammalian species [34]. This also holds true for poultry, as noted by a number of recent studies that establish an important role for vitamins in the maintenance and promotion of a desirable microbiome.

In one of the first investigations [27], the diversity of the cecal bacteria was compared between vitamin-supplemented broilers at NRC [18] levels versus a non-supplemented group in which both groups were fed diets tethered to corn and soybean meal. With vitamin supplementation, the diversity of the microbiome was increased ($P < 0.0001$) with *Clostridium* being the most dominant genus of all species after a 28-day test period, whereas *Faecalibacterium* was second most abundant in those same birds. *Escherichia/Shigella* were found only in broilers without added vitamins in the feed, while *Lactobacillus* was present in those fed vitamins. This work indicated that vitamin supplementation encouraged the presence of advantageous bacteria, whereas the vitamin content supplied only by feed ingredients resulted in a microflora being notably less desirable. In similar work, aged layers benefited with an increased abundance of advantageous ileal and cecal bacteria, along with an improved laying rate and egg quality, when supplemented with a 2-fold higher vitamin level [35]. This work was tethered to the hypothesis that modern laying hen genetics require vitamin levels over and beyond today's standards.

Although it is unclear in these studies which vitamins may have had the biggest impact, recent work with laying hens focused only on vitamin D_3. Laying hens fed a vitamin D_3-deficient diet were particularly susceptible to *Salmonella enterica* and gut mucosal damage [36]. Populations of less favorable intestinal bacteria such as *Escherichia*, *Enterobacteriaceae*, and *Clostridia* were elevated in vitamin D_3-deprived hens, whereas *Lactobacillus* and *Bacilli* and other desirable bacteria became more predominant when vitamin D_3 was supplemented (3000 IU/kg). Hens that were vitamin D_3-deficient experienced mucosal injury and extensive intestinal inflammatory response, leading the investigators to suggest that vitamin D_3 could be an important nutritional strategy to defend against Salmonella infection. More investigation is needed to understand the optimal vitamin D_3 level for this purpose.

In related work, dietary 25-OH vitamin D_3 (HyD®; 69 μg/kg) in laying hens at high stocking density elevated bacterial diversity and improved intestinal function [37]. Villus height was significantly increased with 25-OH D_3 supplementation, as was oxidative capacity. In earlier work, high stocking density elevated blood corticosterone (an indicator of higher stress) and decreased the activities of several antioxidant enzymes [38]. Li et al. [39] reported that 25-OH D_3 improved some oxidant enzyme activities in the small intestine of hens maintained in higher density populations, which implies that the antioxidant effect of 25-OH D_3 might be indirectly improving morphologic features of the intestinal tract. In addition, the microbiota composition was improved with 25-OH vitamin D_3 which included typical vitamin D_3 supplementation levels, suggesting that 25-OH D_3 itself may be important to maintain morphologic and microbiome status.

Broilers challenged with *Salmonella enteritis* also benefited from vitamin C at 500 mg/kg with a reduction in damage to villus structure [40]. Vitamin C supplementation expanded cecal microbial diversity, such as the *Firmicutes* to *Bacteroides* ratio on days 21 and 35. Under heat stress conditions, cecal *Lachnospiraceae* and *Ruminococcacaea* populations were elevated when broilers were supplemented with vitamin E at 250 IU/kg (about 5-fold higher than commercial average), along with an organic Se complex [41]. Both cecal species are important butyrate producers from nonstarch polysaccharides and resistant starch.

As one of the essential nutritional groups, that vitamins can impact intestinal microflora and morphology is not particularly unexpected when considering bacteria also have requirements for certain vitamins and other nutrients. Certainly, these studies offer encouragement that vitamin fortification may include a consideration to account for a healthy and beneficial intestinal microbiome.

8. Egg deposition of vitamins

Partitioning of nutrients by the laying hen generally favors the egg over body tissue. For vitamin E, the yolk is considered the favored tissue for deposition, followed by liver, adipose tissue, dark meat, and white meat [42]. Similarly, the yolk preferentially accumulates folate, as compared to other tissues [43]. The lipid-rich yolk is an important reservoir for the fat-soluble vitamins, as well as for most of their water-soluble counterparts. The exception here appears to be biotin which is found in higher concentrations in the egg albumen [44].

Vitamin	Potential increase	Comments
Vitamin A	2–3-fold	Includes retinol, retinyl esters, and retinal but compiled in yolk mainly as retinol
Vitamin D_3	6–10-fold	Some work demonstrated a much higher accretion in eggs
25-OH D_3	3–4-fold	
Vitamin E	4–5-fold	Considerable variation because of influence of other fat-soluble vitamins
Vitamin K	2–5-fold	Research is limited
Niacin	2–3-fold	U.S. requires niacin fortification to some grain products for human consumption
Thiamin	≈2-fold	Limited data base
Riboflavin	2–3-fold	Refractory to accretion at high feed levels
Pyridoxine	2–3-fold	Limited data base
Pantothenic	2–3-fold	Plateaus quickly
Vitamin B_{12}	3–4-fold	Good response but high feed level needed
Folic acid	2–3-fold	Corn-based feeds may favor higher accretion in yolk
Biotin	3–5-fold	Dietary excess largely goes into albumen

Table 4.
General guideline for egg accumulation of vitamins [14].

Of the commonly accepted vitamins, all but vitamin C are present in the egg. Studies have established a direct link between vitamin levels in the feed consumed and the vitamin concentration of eggs of the hen [45]. Upon absorption, vitamins are delivered to the egg by one of several transport systems that go against a blood/egg concentration gradient. For example, biotin concentration in the egg is about 20-fold higher than in plasma, thus a receptor-mediated transport system works to go from low-to-high gradient [44]. The concentration gradient difference for riboflavin is nearly 7-fold [46], while folate in the yolk was 43-fold higher than in plasma [43].

Yet, owing to the chemical complexity of vitamins, and the absorption and body storage characteristics, as well as the biological variation inherent in hens, the feed-to-egg transfer occurs with a considerable range in efficiency across the vitamins. Whereas vitamin D_3 can be elevated to high levels in the egg through higher supplementation rates [17], other vitamins such as folate are deposited at a much lower rate (**Table 4**) [14]. Various factors such as dietary level, effect of other vitamins and nutrients, period of feeding high vitamin levels, and age of bird can influence the degree to which vitamins accumulate in egg yolk and/or albumen.

9. Commercial broiler vitamin survey

An assessment of vitamin levels used in commercial broiler feeds in the U.S. was recently completed [11], which was conducted in a manner similarly as an earlier study [47]. Supplementation rates were categorized according to feed phase and all vitamins are reported on a pure vitamin level to avoid any bias by differences in product form. Categories included the following: starter (≈day 1–14); grower 1/ Grower 2 (≈day 15–28); finisher (≈day 28–36); withdrawal (WD; ≈day >36); Breeder. Commercial nutritionists provided their addition rates for vitamin premixes which allowed for an accurate calculation for each vitamin per metric ton (MT) of feed. Over 90% of the broiler production for the U.S. was accounted for in this manner.

Poultry feeds are commonly supplemented with vitamins through a vitamin premix that includes fat- and water-soluble synthetic vitamins. In some cases, the same vitamin premix was used across all broiler feeds, or multiple premixes may be fed over the production period. A vitamin premix is typically composed of four fundamental components: vitamins, calcium carbonate (densifier), rice hulls or wheat midds (carrier), and 1–2% oil (to reduce dustiness; adhere vitamins to the carrier). In this survey of commercial vitamin fortification levels, **Figure 3** designates the actual vitamin premix addition rate in terms of kg/MT feed. From starter to WD, the addition rate declined from 0.53 to 0.34 kg/MT, or about 36%.

Figures 4 and **5** illustrate the percent of respective vitamins found in the starter and WD vitamin premixes. These were calculated based on each vitamin in the pure form in the premix. The relative changes between a young bird (starter) and a mature bird (WD) fortification rate are noted in these figures, while those in the grower (not shown) premix ranked between. Niacin, vitamin E and vitamin A were some of the more notable changes as a percent of the total content.

Table 5 lists the average vitamin fortification levels across starter to WD feeds, as well as for broiler breeder feeds. The reduction in vitamin supplementation of feeds from starter to WD parallels the decline in requirements as the bird approaches market age. From starter to finisher, most vitamins declined 20–30%, while from starter to WD, the reduction was greater. The greatest decline occurred with WD folic acid being 52% of the starter level. Several vitamins ranged from 60% to 63% of the

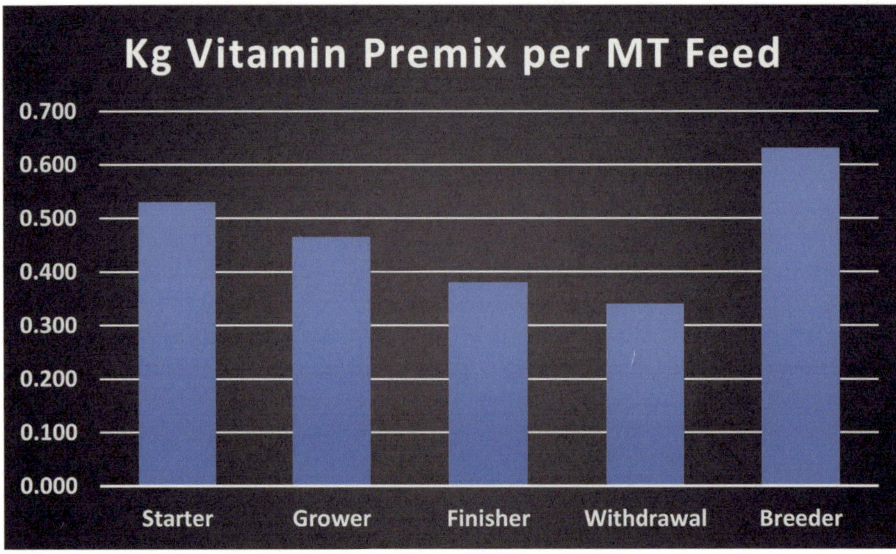

Figure 3.
Vitamin premix addition levels to broiler and breeder feeds [11].

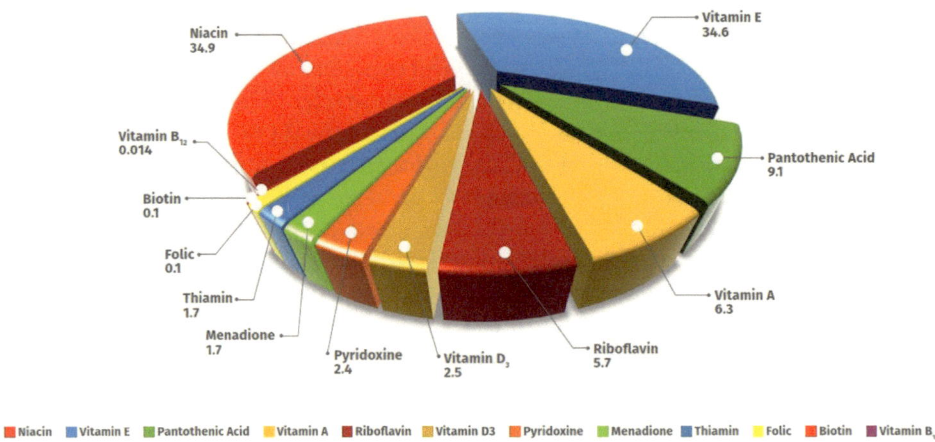

Figure 4.
Broiler starter vitamin premix levels (% of the total active vitamin premix) ([48], unpublished).

level in starter feed compared to WD, and these were vitamin A, vitamin D_3, niacin, pantothenic acid, and vitamin B_{12}.

The 2022 commercial starter vitamin D level was 18.6-fold higher than NRC [18], while vitamin A, E, and K were 6.2-, 5.1-, and 5.0-fold higher in commercial supplementation rates. Differences between commercial water-soluble vitamins and NRC [18] were not nearly as high. Biotin, pyridoxine, pantothenic acid and thiamin difference the least, ranging from 1.2 to 1.4 higher than NRC [18]. The interrelationship among the water-soluble vitamins is recognized as a complicating factor when trying to distinguish individual requirements among that group [49, 50]. Whereas none of the differences between commercial and NRC for the water-soluble vitamin exceeds 3-fold, this is not the case for the fat-soluble vitamins.

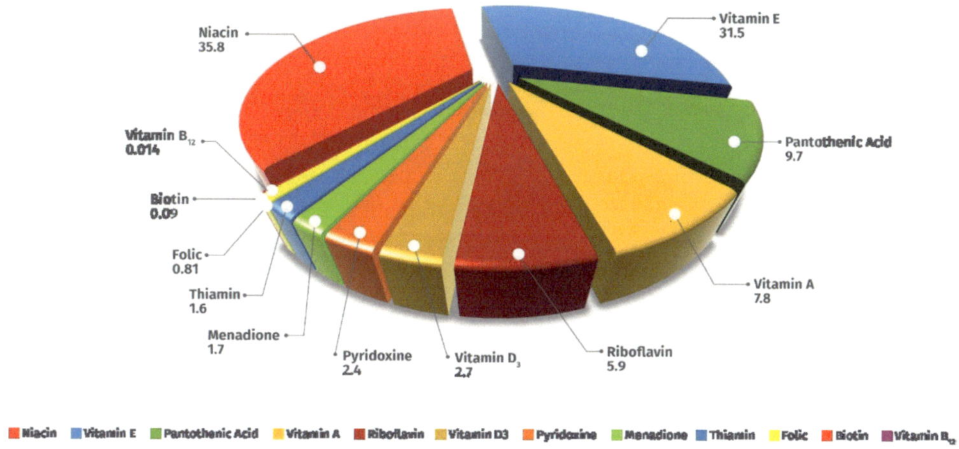

Figure 5.
Broiler WD vitamin premix levels (% of the total active vitamin premix) ([48], unpublished).

The breeder vitamin supplementation premix seldom consists of more of more than one premix over the entire production period (**Table 5**). This survey found that Breeder fortification was higher than typically fed in the starter feeds across all vitamins, presumably because of the need for optimal fertility, egg levels, and chick viability.

Noted in **Table 6** are the overall average values for the starter premix when compared with the average of the lowest 25% and the highest 25% of those values for each of the vitamins. When comparing the lowest 25% average with the highest 25%, vitamins E, K, and B_{12}, along with folic acid and biotin, declined to more than 50% of the

	Type of feed				
Units/MT feed*	**Starter**	**Grower**	**Finisher**	**Withdrawal**	**Breeder**
Vitamin A, MIU	9.29	8.25	7.00	5.73	10.58
Vitamin D_3, MIU	3.73	3.42	2.89	2.35	4.30
Vitamin E, TIU	51.07	41.18	35.15	27.39	67.36
Vitamin K, G	2.52	2.35	2.03	1.50	3.42
Niacin, G	51.60	47.57	41.97	31.08	54.13
Thiamin, G	2.46	2.18	1.87	1.38	3.11
Riboflavin, G	8.49	7.62	6.69	5.15	11.14
Pyridoxine, G	3.57	2.69	2.76	2.12	4.74
Pantothenic, G	13.51	12.70	10.93	8.43	17.31
Vitamin B_{12}, MG	20.56	15.84	15.77	12.57	27.23
Folic, MG	1329.7	1192.9	994.5	696.8	2046.3
Biotin, MG	180.9	131.1	142.2	100.3	267.9

**MIU = million international units; TIU = thousand international units; G = grams; MG = milligrams.*

Table 5.
Average U.S. vitamin fortification rates for broilers and breeders [11].

Units/MT feed*	High 25%	Average	Low 25%
Vitamin A, MIU	10.66	9.29	8.31
Vitamin D_3, MIU	4.74	3.73	2.93
Vitamin E, TIU	72.55	51.07	34.62
Vitamin K, G	3.76	2.52	1.57
Niacin, G	63.30	51.60	41.74
Thiamin, G	3.12	2.46	1.81
Riboflavin, G	9.88	8.49	7.34
Pyridoxine, G	4.55	3.57	2.58
Pantothenic, G	16.65	13.51	10.71
Vitamin B_{12}, MG	31.63	20.56	13.02
Folic, MG	2017.64	1329.66	876.52
Biotin, MG	263.1	180.9	96.0

**MIU = million international units; TIU = thousand international units; G = grams; MG = milligrams.*

Table 6.
Comparison of high and low 25% with average U.S. vitamin fortification rates for broilers and breeders [11].

highest 25%. All others declined to a lesser degree. Overall, biotin declined the most (from highest 25% to lowest 25%) while vitamin A declined the least. Such changes can be attributed to different factors, such as how much agreement there exists within research data for necessary fortification levels, to special effects a vitamin is perceived to have beyond meeting requirements for optimal bird performance.

Figure 6 illustrates the percent coefficient of variation (%CV; standard deviation divided by the mean times 100) by feed phase and within each vitamin category. The lowest %CV across vitamin fortification levels existed within the starter feeds. Vitamin A and riboflavin exhibited the lowest %CV, suggesting the greatest agreement among nutritionists for these two vitamins and nutritional requirements. On the other hand, vitamin E and vitamin B_{12} generally were the most variable across all feed phases, while the remaining vitamins fell someplace within these two groupings.

Listed in **Table 5** is a comparison of vitamin levels from the current 2022 survey and the 1993 survey. In comparing the former versus present, it is worthwhile to mention that 2017–2018 was a period of global vitamin shortages. During this period, many nutritionists were forced to pare back their vitamin supplementation levels in order that all birds received some vitamins, as opposed to some not receiving any supplementation. Current vitamin fortification levels have not fully recuperated to this point and are reflected in the 2022 vitamin summary. Even so, the pressure for more skeleton and body weight in a shorter time requires higher vitamin fortification levels [51, 52].

In reference to the 1993 versus 2022 surveys, the 2022 vitamin levels were higher, especially for vitamin E and biotin, being 184% and 127% of the 1993 levels. Vitamin A, pantothenic acid, niacin, and riboflavin showed less increase, being 5%, 12%, 13%, and 20% higher in 2022 than in 1993. Considering the genetic improvements since 1993, overall increases in vitamin are not necessarily impressive, despite some nutritionists curtailing levels because of shortages.

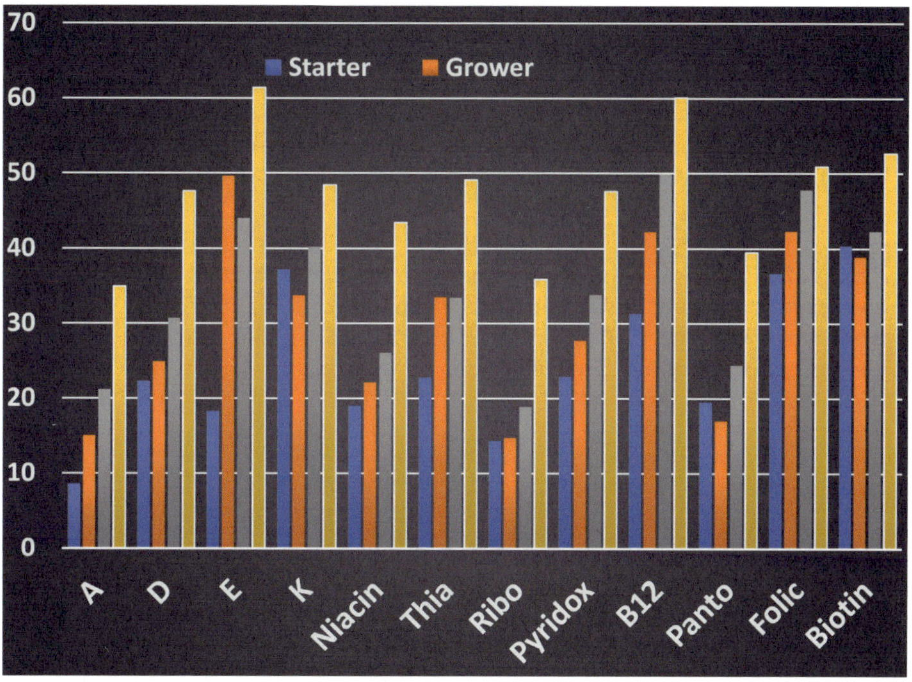

Figure 6.
Coefficient of variation of individual vitamin fortification rates for broilers across feed phases ([48], unpublished).

10. Conclusion

Based on the scientific results presented in this chapter, the following conclusions can be drawn:

1. Vitamins consist of a class of micronutrients that are commonly supplemented to broiler feeds through a vitamin premix.

2. Vitamins represent an important group of nutrients with fortification guidelines that can be affected by stress, genetic change, and other factors.

3. In their natural form in feed ingredients, vitamins vary in levels, bioavailability, and stability in today's feed manufacturing processes.

4. Through chemical and physical means, commercial forms of vitamins are variably stabilized and improved according to their inherent character to withstand environmental challenges.

5. Guidelines for vitamin fortification are developed by various organizations, some of which make allowances for a number of different factors that influence vitamin needs for proper growth and development and return on financial investment.

6. These micronutrients are vital for developing embryos, but not all vitamins are equally transferred from blood to egg, and this should be considered when supplementing breeder feeds.

7. Within recent years, the presence/absence and levels of vitamins have been shown to promote a more desirable microbiome diversity, while defending against *Salmonella enteritis* and improving overall intestinal morphology.

8. Today's vitamin fortification levels exceed NRC [18] guidelines for the fat-soluble vitamins by 5-fold or higher, whereas differences with water-soluble vitamins are also higher but more subdued.

9. Relative to 1993, a similar survey nearly 30 years later showed significant increases for vitamin E and biotin, but less change occurred for other vitamins, while variability in fortification rates were the least variable in starter broiler feeds, and for vitamin A and riboflavin.

Author details

Nelson E. Ward
DSM Nutritional Products, North America

*Address all correspondence to: nelson-e.ward@dsm.com

References

[1] Oviedo-Rondon E, Barroeta AC, Cepero Briz R, Litta G, Hernandez JM. Optimum Vitamin Nutrition for More Sustainable Poultry Farming. Lings UK: 5M Book Ltd; 2023

[2] Ren Z, Yan J, Pan C, Liu Y, Wen H, Yang X, et al. Supplemental nicotinamide dose-dependently regulates body phosphorus excretion via altering type II sodium-phosphate co-transporter expressions in laying hens. The Journal of Nutrition. 2020;**150**:2070-2076. DOI: 10.1093/jn/nxaa148

[3] Savaris V, Broch J, de Souza C, Junior N, de Avila A, Polese C, et al. Effects of vitamin A on carcass and meat quality of broilers. Poultry Science. 2021;**100**:174-185. DOI: 10.1016/j.psj.2021.101490

[4] Shabani S, Mehri M, Shirmohammad F, Sharafi M. Enhancement of sperm quality and fertility-related parameters in Hubbard grandparent rooster fed diets supplemented with soybean lecithin and vitamin E. Poultry Science. 2021;**101**:1-10. DOI: 10.1016/j.psj.2021.101635

[5] Shojadoost B, Yitbarek A, Alizadeh M, Kulkarni R, Astill J, Boodhoo N, et al. Centennial review: effects of vitamins A, D, E and C on the chicken immune system. Poultry Science. 2021;**100**:1-17. DOI: 10.1016/j.psj.2020.12.027

[6] Combs GF, McClung JP. The Vitamins. 6th ed. Cambridge MA: Academic Press with permission from Elsevier Inc.; 2022

[7] McDowell LR. Vitamins in Animal and Human Nutrition. Ames IA: Iowa State University Press; 2000

[8] Semba RD. The long road, rocky road to understanding vitamins. World Review of Nutrition and Dietetics. 2012;**104**:65-105

[9] Carpenter KJ. The Nobel Prize and the discovery of vitamins. 2004. Available from: www.nobelprize.org

[10] Eggersdorfer M, Laudert D, Letinois U, McClymont T, Medlock J, Netscher T, et al. Angewandte Chemie, International Edition. 2012;**51**:12960-12990

[11] Ward NE. Broiler vitamin nutrition guidelines. Ark. Nutr. Conf., Bentonville AR. 2022

[12] Basu TK, Donaldson D. Intestinal absorption in health and disease: micronutrients. Best Practice & Research. Clinical Gastroenterology. 2003;**17**:957-979

[13] Gonclaves A, Roi S, Nowicki M, Dhaussy A, Huertas A, Amiot M-J, et al. Fat-soluble vitamin intestinal absorption: absorption sites in the intestine and interactions for absorption. Food Chemistry. 2015;**172**:155-160

[14] Ward NE. Vitamins in eggs. In: Hester P, editor. Egg Innovations and Strategies for Improvements. Cambridge MA: Academic Press with permission from Elsevier Inc.; 2017. pp. 207-218

[15] Aburto A, Britton WM. Effects of different levels of vitamins A and E on the utilization of cholecalciferol by broiler chickens. Poultry Science. 1998;**77**:570-577

[16] Combs GF. Vitamin tolerance in animals. Washington, D.C: The National Academy of Science; 1990

[17] Yao L, Wang T, Persia M, Horst RL, Higgins M. Effects of vitamin D3-enriched diet on egg yolk vitamin D3 content and yolk quality. Journal of Food Science. 2013;**78**:C178-C183

[18] NRC. Nutrient Requirements of Poultry. Ninth Revised ed.

Washington D.C.: National Academy Press; 1994

[19] Applegate T, Angel RA. Nutrient requirements of poultry publication: history and need for an update. Journal of Applied Poultry Research. 2014;**23**:567-575

[20] Zuidhof MJ, Schneider BL, Carney VL, Korver DR, Robinson FE. Growth, efficiency, and yield of commercial broilers from 1957, 1978, and 2005. Poultry Science. 2017;**93**:2970-2982. DOI: 10.3382/ps.2014-04291

[21] Agri Stats, Inc. Fort Wayne IN. 2022

[22] Patricio IS, Mendes A, Ramos A, Pereira DF. Overview on the performance of Brazilian broilers (1990 to 2009). Brazilian Journal of Poultry Science. 2012;**14**:233-304

[23] Bain MM, Nys Y, Dunn IC. Increasing persistency in lay and stabilizing egg quality in longer laying cycles. What are the challenges? British Poultry Science. 2016;**57**:330-338

[24] Aviagen. Ross broiler nutrition specifications. 2022. Available from: https://e..aviagen.com/brands/ross/

[25] Cobb-Vantress. Cobb 500 nutrition and management guide. 2022. Available from: https://222.cob-vantress.com/resource/product-supplements

[26] McDowell LR, Ward NE. Optimum vitamin nutrition for poultry. Internship in Poultry Production. 2008;**16**:27-33

[27] Luo Y, Peng H, Wright AG, Bai S, Ding X, Zeng Q, et al. Broilers fed dietary vitamins harbor higher diversity of cecal bacteria and higher ratio of *Clostridium*, *Faecalibacterium*, and *Lactobacillus* than broilers with no dietary vitamins revealed by 16S rRNA gene clone libraries. Poultry Science. 2013;**92**:2358-2366. DOI: 10.3382/ps.2012-02935

[28] Liang S, Liu X, Zhao J, Liu R, Huang X, Liu Y, et al. Effects of high dose folic acid on protein metabolism in breast muscle and performance of broilers. Poultry Science. 2022. DOI: 10.1016/j.psj.2022.10935

[29] Fatemi SA, Alqhtani A, Elliott KEC, Bello A, Zhang H, Peebles ED. Effects of the in ovo injection of vitamin D3 and 25-hydroxyvitamin D3 in Ross 708 broilers subsequently fed commercial or calcium and phosphorus-restricted diets. I. Performance, carcass characteristics, and incidence of woody breast myopathy. Poultry Science. 2021. DOI: 10.1016/j.psj.2021.101220

[30] Avila LP, Leiva SF, Abascal-Ponciano GA, Flees JL, Sweeney KM, Wilson JL, et al. Effect of combined maternal and post-hatch dietary 25-hydroxycholecalciferol supplementation on broiler chicken Pectoralis major muscle growth characteristics and satellite cell mitotic activity. Journal of Animal Science. 2022;**100**:1-12

[31] Ward NE. Potential vitamin D role explored in eggs. Feedstuffs. 1 Jun 2009;**81**(22):21-22

[32] Ward NE. Folate-fortified eggs have niche. Feedstuffs. 14 Dec 2009;**81**(51):47-49

[33] Ward NE. Vitamin requirements and economic responses. Multi-State Poultry Conf., Indianapolis IN. May 22-24. 2012

[34] Pham VT, Dold S, Rehman A, Bird JK, Steinert RE. Vitamins, the gut microbiome and gastrointestinal health in humans. Nutrition Research. 2021;**95**:35-53

[35] Gan L, Zhao Y, Mahmood T, Guo Y. Effects of dietary vitamins supplementation level on the production performance and intestinal microbiota of aged laying hens. Poultry Science. 2020a;**99**:3594-3605. DOI: 10.1016/j.psj.2020.04.007

[36] Guo F, Geng Y, Abbas W, Zhen W, Wang S, Huang Y, et al. Vitamin D_3 nutritional status affected guy health of Salmonella-challenged laying hens. Frontiers in Nutrition. 2022;**9**:888580

[37] Wang J, Zhang C, Zhang T, Yan L, Qiu L, Yin H, et al. Dietary 25-hydroxyvitamin D improves intestinal health and microbiota of laying hens under high stocking density. Poultry Science. 2021;**100**:101132. DOI: 10.1016/j.psj.2021.101132

[38] Wang JP, Qiu LY, Gong HJ, Celi P, Yan L, Dig XM, et al. Effect of 25-hydroxycholecalciferol supplementation and high stocking density on performance, egg quality, and tbia quality in laying hens. Poultry Science. 2020;**99**:2608-2615

[39] Li W, Wei F, Xu B, Sun Q, Deng W, Ma H, et al. Effect of stocking density and alpha-lipoic acid on the growth performance, physiological and oxidative stress and immune response of broilers. Asian-Australasian Journal of Animal Sciences. 2019;**32**:1914-1922

[40] Gan L, Fan H, Mahmood T, Guo Y. Dietary supplementation with vitamin C ameliorates the adverse effects of Salmonella Enteritidis-challenge in broilers by shaping intestinal microbiota. Poultry Science. 2020b;**99**:3663-3674. DOI: 10.1016/j.psj.2020.03.062

[41] Calik A, Emami N, Schyns G, White M, Walsh M, Romero F, et al. Influence of dietary vitamin E and selenium supplementation on broilers subjected to heat stress, Part II: oxidative stress, immune response, gut integrity, and intestinal microbiota. Poultry Science. 2022;**101**:101858. DOI: 10.1016/j.psj.2022.101858

[42] Cherian G, Wolfe FW, Sim JS. Dietary oils with added tocopherols: Effects on egg or tissue tocopherols, fatty acids, and oxidative stability. Poultry Science. 1996;**75**:423-431

[43] Sherwood TA, Alphin RL, Saylor WW, White HB. Folate metabolism and deposition in eggs by laying hens. Archives of Biochemistry and Biophysics. 1993;**307**:66-72

[44] White HB, Whitehead CC. Role of avidin and other biotin-binding proteins in the deposition and distribution of biotin in chicken eggs. Discovery of a new biotin-binding protein. Biochemistry Journal. 1987;**241**:677-684

[45] Naber EC. Modifying vitamin composition of eggs: a review. Poultry Science. 1993;**58**:518-528

[46] White HB, Armstrong J, Whitehead CC. Riboflavin-binding protein. Concentration and fractional saturation in chicken eggs as a function of dietary riboflavin. Biochemical Journal. 1986;**238**:671-675

[47] Ward NE. Vitamin supplementation rates for U.S. commercial broilers, turkeys, and layers. Journal of Applied Poultry Research. 1993;**2**:286-296

[48] Ward NE. Vitamin survey of U.S. broiler industry. Unpublished. 2023

[49] Bains B. A Guide to the Application of Vitmains in Commercial Poultry Feed. Australia: Rath Design Communications; 1999

[50] Whitehead CC. Requirement for vitamins. In: Fisher C, Boorman KN, editors. Nutrient Requirements of

Poultry and Nutritional Research. Oxford, UK: Butterworths; 1987

[51] Jiang RR, Zhao GP, Chen JL, Zheng MQ, Zhao JP, Li P, et al. Effect of dietary supplemental nicotinic acid on growth performance, carcass characteristics and meat quality in three genotypes of chicken. Journal of Animal Physiology and Animal Nutrition. 2011;**95**:137-145

[52] Mejia L, Turner B, Ward N, Burnham D, DeBeer M. Evaluating the effects of feeding OVN and US industry fed vitamin levels on growth performance, processing yields and profitability. T140. IPSF, Atlanta GA. 2014

Chapter 3

Role of Feed Additives in Poultry Feeding under Marginal Environmental Conditions

Ahmed El-Baz and Raafat Khidr

Abstract

Modern commercial breeding programs aim at maximizing productive performance, especially with modern broiler chicken strains, which are characterized by a high growth rate and a lower feed conversion factor. However, it is more sensitive to environmental stress, intensive rearing conditions, and high nutritional needs. Nutrition plays a key role in achieving the maximum amount of production while maintaining the health of the bird, in addition to reducing production costs by searching for unconventional feed ingredients or using some feed additives. Feed additives are mainly used in animal feed to help provide for the bird's needs. In addition, it is used to enhance bird health, stimulate digestion, improve feed efficiency, and resistance to diseases by positive influence on the gastrointestinal tract, metabolism, and enhancement of the immune system, inhibiting pathogens, and improving intestinal integrity. In broiler nutrition, special attention should be paid to feed additives of modification of immunity and microbial content such as pre-and probiotics, nano compounds, herbs. In this chapter, we will elucidate the importance of feed additives from the point of marginal environmental conditions, which face many challenges concerning poultry feeding. Hence, feed additives will be a fruitful tool to cope with some of such challenges under those marginal conditions.

Keywords: poultry, feed additives, performance, heat stress, antibiotics, unconventional feed

1. Introduction

The poultry industry is one of the largest investments in the world. In particular, broiler chicken production, which attracted enormous investments for business profit, as it is characterized by low production costs and short production period. In addition to the increased demand for poultry meat due to its low-fat content and low consumer price compared to other meats (animals and fish). The consumption of poultry meat represents about 70% of the total meat consumed as about 66 billion birds are slaughtered annually, and the United States, China, and Brazil represent the largest poultry-producing countries. With this remarkable development in the broiler industry, it was necessary to use some feed additives to meet many challenges, including disease resistance, prevention of heat stress, improving the utilization of feed, and stimulating growth and production [1–3]. In this concern, feed additives are products used for specific purposes in animal feed, to meet the poultry nutritional requirements and improve the quality of feed, and enhance the animals performance and health, as well as the quality of food of animal origin (e.g., eggs and meat) [4, 5].

Global warming is one of the major challenges for animal breeding. High environmental temperatures negatively affect the poultry industry; thus, we have a fundamental interest in reducing the negative effects of climate change on poultry breeding. The important question is, what tools do we have to reduce the harmful impacts of high environmental temperatures? A solution for the prevention of heat stress in poultry includes developing technology devices (e.g., air conditioning and intensive ventilation); however, housing methods are more expensive and this makes it the biggest obstacle to its spread, especially in developing countries. Therefore, reducing the harmful effects of heat stress with different nutritional tools such as using a feed additive [6, 7]. Before selecting additions, we must be aware of the changes in the physiological and metabolism of broilers caused by heat stress to determine the type and role of each feed additive that can be used. Numerous studies have shown that the use of some feed additives (such as plant extract, probiotics, vitamins C, E, and A, zinc, and selenium) had a positive role in mitigating the harmful effects of heat stress [7–9].

It was reported that the use of antibiotics is necessary to fight pathogenic microbes, in particular, infectious pathologies (*Clostridia* and *Coccidiosis*), as well as growth promoters (regulation of the intestinal microflora, increased vitality, enhancement growth performance, and stimulation of the immune system) [10, 11]. Despite all these advantages, but as a result of the wrong use of antibiotics, led to the emergence of antibiotic-resistant bacterial strains with residues in animal meat, which affects human health. For this reason, the European Union in 2005 banned the use of antibiotics as a growth promoter to minimize health risks [12]. Nutritionists began searching for safe alternatives to antibiotics for animals and humans. In this concern, the use of some feed additives reduces the problems faced by the poultry sector, such as heat stress, and improves feed utilization, subsequently reducing production costs [13–16]. Therefore, this chapter aims to summarize how various feed additives can reduce the use of antibiotics, the negative effects of heat stress, and improve the utilization of unconventional feed ingredients.

2. Using feed additives to eliminate harmful effects of heat stress on poultry

Heat stress is considered a critical holdback to coping with the poultry industry, particularly in a hot environment, triggering greater economic losses in the poultry industry [8, 17]. The bird begins to experience heat stress when the ambient temperature elevation is above the comfort zone (thermoneutral zone) leading to inducing stressful behavioral responses (higher respiratory rate, disorders in metabolism, and *injuries intestinal* integrity and morphology) to loss of excess heat that exceeds the critical temperature (**Figure 1**) [18, 19]. This means that there is more loss of heat (energy), which leads to less energy remaining for production (growth and egg production) and results in poultry performance deteriorating such as dehydration,

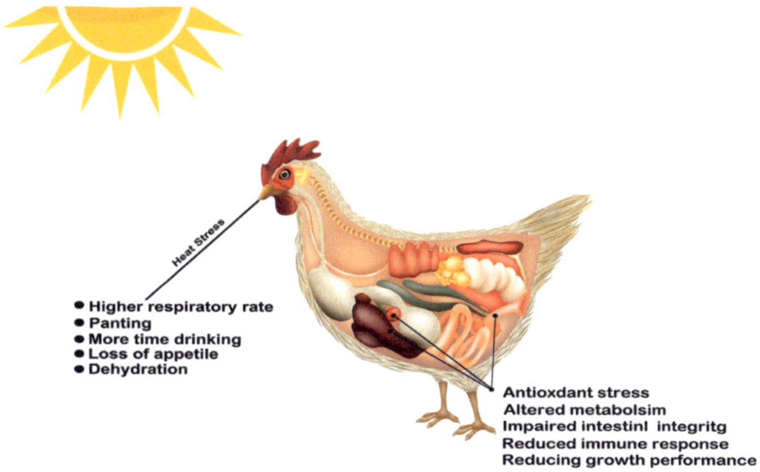

Figure 1.
Physiological, behavioral, and growth performance changes during exposure to heat stress.

high death, and altered meat quality characterized [6, 8]. Furthermore, developments in the genetic selection of broilers have led to rapid growth with a low feed conversion ratio and a high metabolic rate, which is accompanied by increased feed intake and causes a higher heat production level, this made him more sensitive to a hot environment [7]. Summarizing, it can be difficult to keep poultry in a thermoneutral zone in a hot environment, wherefore, it is important to use technical devices (ventilation and a cooling system), in addition to nutritional tools to reduce the negative effect of heat stress on birds. Therefore, the proper management practices can be complemented to keep health and performance by using some vital dietary doctrinaire (feed additives), so phytobiotics, probiotics, natural antioxidants, vitamins, electrolytes, and fat (**Table 1** and **Figure 2**), in addition, to feed form, feed restriction, and drinking cold water [6, 7, 25].

2.1 Oil or fats

Many studies indicated that the addition of fats in chicken diets enhanced the performance index, which was bred under high ambient temperatures [26, 27]. In general, energy is a limiting factor for high ambient temperature conditions, where a deficiency in energy intake could occur through reduced feed intake and appetite. Reduced feed intake during heat-stressed chickens causes the intake of protein and fat to be about 40% less than the needs required to maintain life and productivity [26], in addition to minimizing the heat production induced by feed digestion. Furthermore, chickens resort to panting for getting rid of the thermal burden, which depends on respiratory muscular activity, resulting in increased energy expenditure [28]. Therefore, chickens need to increase energy while obtaining an easy source to benefit from, for this reason, it was better to provide the required energy during the period of heat stress by adding oil or fats than using carbohydrates [8, 29]. Adding oil contributes to reducing heat production because the heat increment of carbohydrates or protein is higher than that of fat [30]. A significant improvement was observed in the performance and digestion coefficient of birds that received higher recommended

Additives	Findings	Reference
Selenium	Enhancing thyroid hormone metabolism, immunity, and antioxidative	[20]
Nano-Selenium	• Improvements in feed conversion ratio, weight gain, and feed intake. Peroxidase mRNA expression in liver. • Decreasing liver and breast muscle contents of malondialdehyde.	[6]
Zinc	• Improving growth rate, FCR, and nutrient digestibility • Enhancing antioxidant enzymes and humoral immune response	[21]
Zinc, or magnesium, their combinations	• Improving the quails performance • Enhancing humoral immunity • Decreasing of meat lipid peroxidation	[22]
Vitamin C	• Improving the performance status. • Enhancing immunological traits, and behavior	[23]
Probiotic	• Improving bacterial population of the cecal contents, and immune response • Improved daily weight gain and decreased mortality rate	[16]
Spirulina platensis as probiotic	• Enhancing the productive performance • Enhancing the lipid profile, redox status, and humoral • immune response	[7]
Probiotic, citric acid, garlic powder or their combinations	• Improving of body weight gain, feed conversion ratio, and reduction in abdominal fat. • Enhancing the feed utilization by highest crypt depth values • Improving of immune system by highest antibody level against NDV	[3]
Betaine	• Better feed conversion ratio and performance efficiency factor. • Enhancing meat characteristics by improve breast fillets yield.	[24]
Herbs	• Improving the immune system and antioxidative status • Stimulating digestive enzyme activity, and control pathogenic bacteria	[16]
Essential oil mixture (garlic and lemon)	• Increasing digestive enzymes activities, and improving average body weight, feed conversion ratio • Enhancing intestinal microbial content, and intestinal histological status	[7]

Table 1.
Using feed additives to eliminate harmful effects of heat stress on poultry.

levels of metabolizable energy during heat stress. Al-Harthi et al. [26] declared that heat-stressed broilers fed a high metabolizable energy diet showed an improvement in feed utilization and live body weight. Adding oil contributes to reducing heat production because the heat increment of carbohydrates or protein is higher than that of fat [30]. Moreover, fat-supplemented diets improved the palatability of poultry diets and physical characteristics, resulting in enhanced feed intake and performance

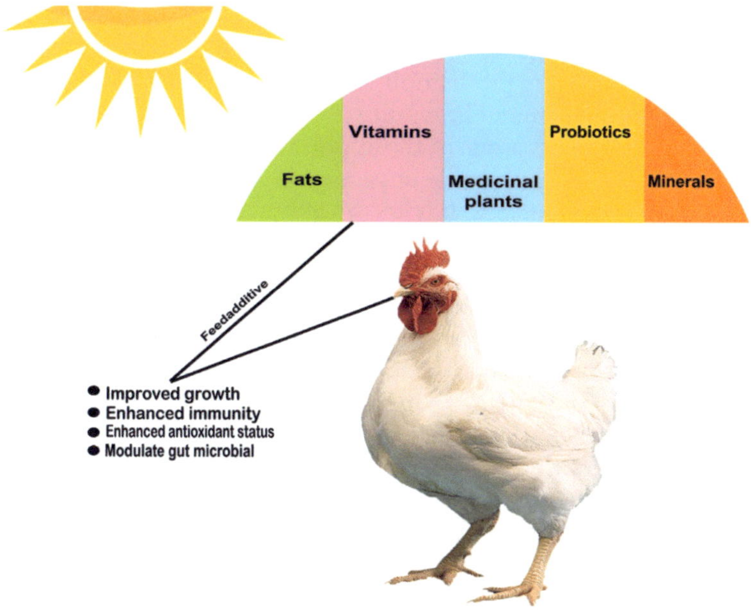

Figure 2.
The role of dietary supplements in mitigating the adverse effects of heat stress.

[31]. In addition, many studies confirmed that oil supplementation in the poultry diet increases nutrient utilization by lowering the rate of food passage in the gut [32].

2.2 Minerals and vitamin

Minerals and vitamins play an important role in all vital processes in the body, including the basal metabolic rate, antioxidative properties, and protein composition, resulting in improved health and performance.

Selenium (Se) is one of the important elements that acts as a cofactor for antioxidant enzymes, as well as responsible for the conversion of thyroxin (T4) into active triiodothyronine (T3) and is involved in several biological functions. Several studies indicated that dietary supplementation of Se or Nano-Se enhanced BWG, FCR, and immune responsiveness in heat-stressed broilers [6, 33, 34]. Moreover, numerous studies have shown that adding Se to chicken diets reduces the negative effects of heat stress by enhancing thyroid hormone metabolism, immunity, and antioxidative status (elevated mRNA expression of GSH-Px in the liver), which resulted in improved productive performance [20, 34].

Zinc (Zn) is an essential element that contributes to bone formation, feather formation, and composition, and the function of more hundred enzymes linked to the metabolism of nucleic acid, energy, and protein [35]. Moreover, Zn is involved in the activity of the antioxidative enzymes (GSH-Px, SOD) by suppressing free radicals [36]. Numerous research also confirmed that the addition of zinc in its various forms (organic, inorganic, and nano) in chicken feed exposed to environmental heat stress led to improved growth rate, FCR, nutrient digestibility, minimized lipid peroxidation, and enhanced antioxidant enzymes and humoral immune response [21, 37]. Recently, mixing two or more elements had a positive effect in reducing the harmful effects of heat stress, such as mixing between magnesium (Mg) and zinc(Zn) diet increased BWG and dressing percentage in Japanese quail [22]. Moreover, the addition of an organic minerals mixture (Zn, Cu, and Mn) in heat-stressed laying hen feed, resulted in improved egg production, egg quality traits, minimized yolk lipid oxidation, and enhanced humoral immunity [37, 38].

Ascorbic acid (vitamin C) is one of the important antioxidant components (water-soluble) that attenuate the undesirable impacts of heat stress in poultry *via* safeguards cells against oxidative damage [39]. It is known as ascorbic acid and is endogenously synthesized in several bird species, however, necessary to be adding to the diet under conditions of heat stress. This can be explained by exposure to high ambient temperatures reducing the absorption of vitamin C and accelerating its destruction. Vitamin C is synthesized in poultry in the kidney from glucose, but in normal conditions, the birds are able to synthesize adequate amounts of vitamin C. A study showed a significant decrease in the level of ascorbic acid by 40% in the blood of Japanese quails under heat stress conditions compared with the control group [40], thus the increased requirements of this vitamin C during periods of heat stress [41]. Beside, vitamin C plays a role in improving feed efficiency by stimulating the thyroid gland, as well as enhancing Ca + 2 metabolism *via* participation in essential processes such as adrenaline, corticosterone release, and 1, 25-dihydroxy vitamin D biosynthesis [23]. In addition, it is necessary for immune system activation and body temperature regulation. Because of the increased intensity of heat stress on birds performance, many scientific studies that vitamin C supplementation in dietary stressed Japanese quails leads to improving the performance status, enhancing immunological traits, and behavior, and declines the metabolic rate and survival rate [25, 42]. Additionally, supplementation of the vitamin C diet decreased lipid peroxidation and improved the antioxidant status in Japanese quail exposed to heat stress [42].

2.3 Direct-fed microbials

Heat stress results in the impairment of gut integrity and function by impairing intestinal microflora balance, and mucus layer [43, 44]. This disruption of the intestinal barrier facilitates the translocation of pathogenic bacteria and their toxins into the host body (bird) and enhances inflammatory responses. Previous studies noted that acute heat stress leads to a significantly altered gut microbial community (increased opportunity for *Salmonella* attachment) and intestinal morphology [43, 45]. The strategy aims to use probiotics as a means to mitigate the negative effects of heat stress on the bird by modifying the microbial content to optimize gastrointestinal health. Probiotics or direct-fed microbials are microbial and are defined as live beneficial microbial feed supplements, including bacteria (*Bifidobacteria, Bacilli,* and *Lactobacilli*), fungi (*Aspergillus awamori*, and *Aspergillus oryzae*), and yeast cultures (*Saccharomyces*) that can intestinal microbial balance, intestinal health, and immune responses, results in improve poultry performance [3, 46]. Several studies summarized that probiotic supplemented in broiler diets improved daily weight gain and decreased mortality rate under hot environmental conditions [8, 47]. The inclusion of probiotics in broiler diets leads to enhanced growth performance, and immune response [16]. Therefore, it is extremely useful to equip poultry diets with probiotics levels that support intestinal integrity and improve feed utilization by increasing intestinal absorption surface by increasing the height of the villus (**Figure 3**) and enhancing the immune system. As indicated by a study by Abdel-Moneim et al., [15]; Elbaz et al., [3]; Li et al., [48] that the inclusion of *Bacillus subtilis* has the potential to produce some digestive enzymes, as well as, enhance

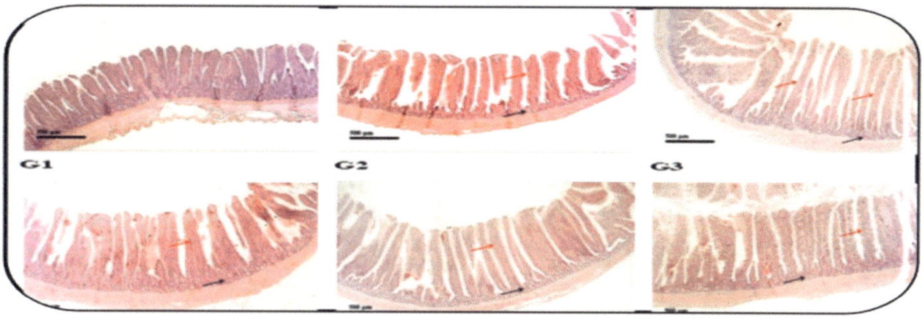

Figure 3.
Photomicrographs of ileal villi showing how villus height of broiler chickens exposed to heat stress (G1) is shorter compared to other broiler-fed on some feed additive.

intestinal development and function, and improve immune response. Moreover, Abdel-Moneim et al. [15]; Saleh et al., [49] indicated supplementation of *B. subtilis* in quail and broiler diets also improvement in feed efficiency and BWG, and raise muscle concentrations of unsaturated fatty acids. Many studies have also confirmed the important physiological role of probiotics in promoting the antioxidant defense system of heat-stressed broiler chickens via activating enzymes antioxidants [6]. For this reason, the positive role of adding probiotics can be emphasized in mitigating the negative effects of heat stress via enhancing intestinal integrity and improving feed utilization and antioxidant status.

2.4 Medicinal plants and plant bioactives

Herbs, plant extract, and essential oils spice is widely used in herbal medicine to improve the immune system and enhance antioxidative status and antimicrobial. Volatile essential oils can stimulate digestive enzyme activity and control pathogenic bacteria [7, 50]. Heat stress disrupts the balance between oxidation and antioxidant defense systems, causing lipid peroxidation, and consequently DNA oxidative damage [33]. Plant bioactives are a type of chemical found in plants and certain foods with small amounts (such as vegetables, fruits, oils, and grains). Essential oils have the biological activity of substances with different chemical compositions and concentrations, which play of important in antioxidant activity, anti-inflammatory, and antimicrobial (thymol and carvacrol) [51, 52]. In addition, essential oils or organic acids, including lactic, citric, formic, and fumaric acids, generated from plants are to control harmful microorganisms in the digestive and respiratory organs of poultry *via* reducing pH in the gut [53]. Results of many research indicate that supplementation of thyme essential oil is a suitable strategy to improve the immune system and productive performance, decreased the mortality rate, and reduce the negative effects of heat stress [7, 54, 55]. Supplementation of essential oil broilers diet had reduced the adverse effects of heat stress on performance and immune responses [55] and can be a good alternative to improve the adverse impact of aflatoxin B1 contaminated in the broiler diet [56].

2.5 Betaine

Betaine is the amino acid glycine or trimethyl glycine and is greatly found in a variety of plants. Functionally, betaine plays an important role in mitigating the harmful effects of heat stress through its role as a methyl group donor for the methionine homocysteine cycle and as an organic osmolyte, as well as it is an antioxidant natural. It also plays an important role under conditions that inactivate cells (loss of water causes cells to die) *via* protecting cells from osmotic pressure, which allows them to continue normal metabolic activities. This confirms the evidence that biotin may be a feed material with positive effects on poultry performance, especially during heat stress through the high value [24, 57]. Many studies have confirmed that the addition of biotin improves broiler performance, carcass composition by changing lipid metabolism [58], immune response, lipid metabolism [59], and intestinal barrier function [60]. Furthermore, it has been suggested that improving the meat quality of broilers fed on betaine is due to its role as a natural antioxidant [61].

3. Use of feed additives as alternatives to antibiotics in poultry diets

For many decades, poultry breeders have been looking for a growth stimulant and protection for birds from intestinal and respiratory diseases, even using antibiotics. Several previous studies have shown that the addition of antibiotics in poultry feed improves productive performance by stimulating the immune system, increasing vitality and regulation of the intestinal microflora, and improving appetite and feed conversion efficiency [11, 62]. Despite all these desirable advantages, the incorrect use of antibiotics has led to increasing antimicrobial resistance bacteria and residues in animal products as a public health threat. For this reason, the European Union in 2005 banned the use of antibiotics as growth promoters. This motivated nutritionists to search for safe alternatives to antibiotics to maintain public health while increasing antibiotic-free broiler meat production. Therefore, this chapter part aimed to explain feeding strategies of different antibiotic alternatives, including prebiotics, probiotics, enzymes, and phytogenic groups (herbs, essential oils, and marine algae), [2, 3, 13, 16] in poultry production (**Table 2** and **Figure 4**). Will be explained in the following points.

3.1 Probiotics

Probiotics are live strains of beneficial bacteria that confer a health benefit on the host by fighting pathogens in the gastrointestinal tract of chickens, enhancing immunity, and stimulating growth. In addition, it is providing feeding efficiency improvement, antioxidant capacity, the microbial profile of the cecum, and intestinal protection [5, 15]. Several strains of beneficial microbes have been identified in the bird's gut for development and use as probiotics [3, 15]. The most used microorganisms as probiotics in poultry feed are bacterial strains (Gram-positive) such as *Lactobacillus*, *Bifidobacterium*, *Bacillus*, yeast (*Saccharomyces*), and fungi (*Aspergillus*). The main action of probiotics includes lowering the gut pH through the organic acids and volatile fatty acids produced during the fermentation process through probiotics [16]. As well low pH in the intestine inhibits the colonization effects of pathogens in the digestive tract. Probiotics work as well by secreting products that inhibit their development such as organic acids, bacteriocins, and hydrogen peroxide and competitive exclusion through competing with pathogenic bacteria for locations in the intestinal mucous membrane to adhere to nutrients [16, 68]. Other principal

Additives	Type	Findings	Reference
Probiotic	Bifidobacterium	• Increased body weight and weight gain and enhanced feed conversion ratio • Improved antioxidant status and immune response • Improved Ileal architecture by highest values of villus height	[5]
	Multi-strain probiotic	• Improved broiler growth performance • Reduced ileal enumeration of *E. coli* and total coliform and increased *Lactobacillus* count • Decreased abdominal fat and no effect on carcass weight • Enhanced immune response	[3]
	B. subtilis	• Improved live body weight and feed-to-gain ratio • Increased serum total protein and albumin levels • Triiodothyronine and thyroxine activities were significantly elevated • Promoted the antioxidative status and digestive enzymes activities	[5]
Prebiotics	*S. cerevisiae*	• Enhanced feed efficiency and performance • Improved gut morphological structure and reduced the number of pathogenic bacteria • Stimulation of the host adaptive immune system	[63]
	Mannan oligosaccharides	• Improved growth performance and intestinal oxidative status • Increased the relative weight of the bursa of Fabricius and jejunal immunoglobulin content, • Decreased cecal Salmonella colonies	[64]
Organic acids	Citric acid	• Improved growth performance and decreased abdominal fat • Decreased serum concentrations of cholesterol, triglycerides and LDL, while HDL was elevated • Modification of the microbial content	[3]
	Formic acid	• Increased body weight gains and decreased feed consumption • Increased total fat and bursa weight • Decreased coliform counts and the pH	[65]
	Butyric acid	• Improved body weight gain and feed conversion ratio • Enhanced AME_N content and apparent ileal digestibility of amino acid	[66]
Phytogenic	Thyme oil	• Better feed conversion ratio • Decreased the plasma AST and increased total protein • Increased antibody titer to infectious bursa disease	[54]

Additives	Type	Findings	Reference
	Curcumin	• Enhanced performance and health status • Better carcass traits and decreased abdominal lipids • Improved Meat quality and cecal microbial counts	[67]

Table 2.
Use of feed additives as alternatives to antibiotics in poultry diets.

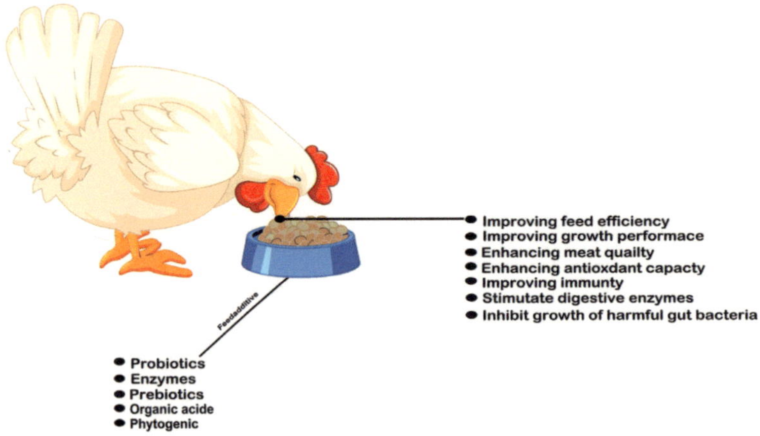

Figure 4.
The most important antibiotics alternative to enhance the general performance of chickens.

mechanisms of probiotics are also used to modulate immunomodulation and to improve intestinal integrity by modulating intestinal microbiota and competition for binding sites on the intestinal epithelium wall, which hinders competition and joining of pathogenic microorganisms, this higher concentration of the beneficial microbiota. The results of several studies showed that feeding poultry on probiotics enhanced fiber and protein digestion and enzymatic activity, resulting in efficient feed nutrient utilization [67, 69]. Some studies also confirmed that adding a mixture of beneficial microbes (*B. subtilis and Lactobacillus*) was more effective in performance in environmental stress conditions through promoting nutrient digestion and gut health and the immunity modulated by the microbiota [68, 69].

3.2 Prebiotics

Prebiotics are carbohydrates that can be utilized by useful gut microorganisms but are indigestible by the host animal. The most important sources of prebiotics are mannan oligosaccharides, fructooligosaccharide, oligofructose, inulin, galactan, galactooligosaccharides, and fiber components, which can extract from barley, oats, flax seeds, onion, and garlic, as well as *green algae*.

Feeding chickens on a diet containing prebiotics have been shown to gut microbiota modulation and improve immunity, which are antioxidant, and antibacterial properties [70]. The most common commercial feed nutrients in poultry feed are mannan oligosaccharides, D-mannose, and β-glucan, which are derived from the cell wall of *Saccharomyces cerevisiae* [63]. The positive role of adding probiotics can be explained by alterations of gut microorganisms that enable them to reduce pathogenic bacteria, increase their numbers of beneficial bacteria, maintain optimal intestinal pH, increase nutrients digestibility, and increase mineral and vitamin absorbability, which improved host health and performance [70, 71]. Modifying the intestine microbiota by promoting beneficial gut microbes that ferment them, leads to the production of short-chain fatty acids, or some antibacterial substances such as bacteriocins against pathogenic microbes [70]. These fermented products of beneficial microbes due to improving the integrity of intestinal epithelial cells [72] will be followed by increasing the absorption of nutrients and improving the growth performance of poultry. Numerous studies have demonstrated the effective role of establishing a healthy microbial community in the intestine of poultry by enhancing the abundance of *Bifidobacteria* and *Lactobacilli* and reducing harmful microbes such as coliform [64, 73]. In addition to stimulation of the immune system, improvement of the epithelium by regulation of the interaction among the host (birds) and the intestinal microbiota, thus improving the productive performance of poultry.

3.3 Organic acids

Organic acids (OC) are organic compounds with acidic properties (weak acids), classified based on the number of carboxylic acid groups (R-COOH) and antimicrobial effects in animal feeds. Organic acids are promising alternatives. Among these, are formic acid, citric acid, propionic acid, and acetic acid. Their inclusion in poultry feed has been shown to enhance growth and feed efficiency [74, 75]. Organic acids individually or in their combinations are usually considered safe and perform can function similarly to antibiotics [76], furthermore, used as feed preservatives due to their strong antifungal and antibacterial properties [74].

The antimicrobial mechanism of organic acid has been suggested for the lowering of the pH of the intestines, that way limiting the growth of the microbial less tolerant to acid pH. The magnitude of the antimicrobial activity of an acid depends upon its concentration and pH [3, 77]. The use of OC has been reported to protect poultry by competitive exclusion [70], and it can penetrate the bacteria cell wall and disrupt the normal physiology of certain types of bacteria, mostly pathogenic microbes (pathogenic bacteria reside at a pH close to 7) [65, 75, 77]. In addition to the antimicrobial activity, they reduce the pH of digesta, increase pancreatic secretion, and prevent damage to epithelial cells by reducing the production of toxic components *via* the bacteria and colonization of pathogens on the intestinal wall, thus enhancement of nutrient utilization and growth and feed conversion efficiency [3, 74]. Several studies have documented the positive effect of OC on improved growth and enhanced digestibility of nutrients and gut health by improved duodenal villus height, boost gastric proteolytic activity (activating the pepsin activity), enhanced absorption of the feed contents from the intestines, and the digestibility of minerals. Furthermore, an increase in antibody titer against Newcastle disease, and improved immunoglobulin status were significantly improved in broilers fed on organic acid supplementation [65, 78]. Similarly, the improved immune response has also been reported in response to organic acid supplementation in broilers could be due to the increased Lactobacillus spp. population in the gut, which has a positive effect on the host immune system [3, 65]. For that can the use of organic acids as a sustainable and potent alternative to antibiotics, thus maximizing future production and health of poultry.

3.4 Phytogenic

Phytogenic are plant-origin extracted compounds that include herbs, spices, and essential oils. It also features less toxic, residue-free, and perfect feed additives for poultry production compared with synthetic antibiotics. Herbs also contain essential oils, organic acids, and a complex mixture of various compounds. Essential oils have many biological properties such as antimicrobial antioxidant enzymatic, anti-heat stress effects, activating the immune system, and stimulating digestion [7, 79, 80]. Organic acids, produced by plants, are which play an important role in controlling harmful microorganisms in poultry's digestive and respiratory organs. The most important of them are lactic, citric, formic, and fumaric acids. The most critical role of organic acids is to reduce pH in the gastrointestinal tract and enhanced the immune responses of poultry [79]. Additionally, organic acids can preserve the microbial balance in the gastrointestinal tract by inhibiting microorganism growth in food and the gut. Many active components (flavonoids, hydrolyzable tannins, proanthocyanidins, phenolic acids, and phenolic terpenes) can prevent lipid peroxidation by the activation of antioxidant enzymes (glutathione reductase, superoxide dismutase, and catalase) or quenching free radicals [80, 81].

Numerous studies have shown that adding herbs or essential oils to chicken feed improves anti-oxidative and antimicrobial activities, reduces inflammation, enhances intestinal functions, and increases fiber and nitrogen retention digestibility [79, 82], which results in improving growth performance. Previous reports confirmed beneficial effects to improve performance and broiler health, which can be used as a good alternative to antibiotics.

4. Using feed additives to improve the utilization of some unconventional feed ingredients in poultry feeding

The major constraint in poultry feeding is the higher prices of conventional feedstuffs (mainly, corn and soybean), which are transported from many countries adding to the cost of production to a great extent. Moreover, these two feed ingredients are also high in demand by other humans (yellow maize) and animals (soybean meal). As a result, the availability of feed ingredients for poultry feed would become more competitive. Feed is one of the major constituents in poultry production, which represents about 80–85% of the total cost of poultry production. This caused an increase in the responsibility of nutritionists increasing the poultry production and research utilization of unconventional feed resources through strategic and applied research to bring down the cost of production. In addition, recently, corn has been used as a major source of produce biofuel, and this further poses a serious food security risk, especially in developing countries [83]. Currently, efforts are large to use alternative sources of energy and protein to be substituted for corn and soybean meal in monogastric animals [46, 84]. As produced a huge amount of alternative feedstuffs in some developing countries are considered as agro waste by-products such as cotton seed meal, olive cake, wheat bran, rice bran, canola meal, palm kernel cake, etc. [46, 85–87]. However, many of this agro-waste products are containing the presence of non-starch polysaccharides (NSPs) such as cellulose, hemicellulose, and lignin, as well as anti-nutritional factors [46, 87], which can negatively effect on productivity and health status of the chickens. Poultry is monogastric animals that lack fiber-degrading enzymes for the breakdown of complex carbohydrates [88].

So, there is a need to improve the utilization of these fibrous materials (unconventional feedstuffs) to incorporate such ingredients in poultry feed without any adverse effect on their health and production. In the following, we will throw some light on to use of some nonconventional feedstuffs with the potential to be replaced partially or totally with corn and soybean meal in poultry feeds, in addition to the importance of some additives to alleviate the secondary metabolites in such feedstuffs (**Table 3**).

4.1 Canola meal

Canola meal (CM) is the by-product of oil extraction. It has a higher crude protein content of approximately 35–40% and sulfur-containing amino acids are higher than that of soybean meal, while lysine content is less than that of soybean meal. The problem with using CM in poultry feeds is the containing of glucosinolate, fibers, sinapine, tannins, and phytate, as well as it has low metabolizable energy [79, 86]. Many methods help to reduce anti-nutritional factors, one of these methods is adding some feed additive, fermentation, etc. [79, 88]. Many studies have shown that adding probiotics, extrusion, exogenous enzymes, or using the fermentation process for some dietary ingredients has improved performance, increased nutrient digestibility, and reduced the effects of antinutritional [96, 97]. Therefore, it was reported that canola meal can be incorporated in poultry diets up to 5–8% without any feed additive, and broiler chickens were fed on a diet containing 20% of fermented CM, which did not negatively affect performance [84]. The fermentation process leads to reducing pathogens such as *Escherichia coli*, and *Clostridium perfringens*, resulting in enhanced gut health [85, 96]. Furthermore, the addition of exogenous enzymes is important to degrade complex fibers (non-starch polysaccharides, NSP) to improve the nutritional value of unconventional feed ingredients [85]. Enzymes play an influential role in improving feed digestion and utilization. Moreover, exogenous enzymes improved nutrient digestibility in poultry leading to lowering nutrient excretion in excreta such as excess nitrogen, phosphorus, and zinc, which reduces environmental pollution and improves feed utilization, in addition, reduced the effect of anti-nutrients and improved productive performance. Previous studies indicated that the fermentation broiler feed by Aspergillus resulted in an increase in nutrient solubility and digestibility, reduced phosphorus excretion, and improved broiler growth and feed utilization compared to the control group [89]. However, positive effects were observed when the addition of enzymes in broiler diets containing canola meal (17.5%) on the overall performance of broilers [95]. It was reported that CM can be incorporated in poultry diets up to 20–25% fermented CM-based with exogenous enzymes in broilers fed [46, 85].

4.2 Sunflower meal

Sunflower meal, a by-product from the oil extraction industry, is available in significantly high quantities throughout the year at a lower cost than soybean. Sunflower meal has protein levels ranging between 30 and 37%, it is a good source of protein with amino acid availabilities similar to those of soybean meal [95]. One of the important characteristics of sunflower meal is that it does not have anti-nutritional factors like those found in soybean. Despite this, its addition to poultry feed does not exceed 15% because it contains a high concentration of non-starch polysaccharides, in addition to low metabolizable energy and lysine levels [98, 99]. Some studies recommended the sunflower meal up to 15% in broiler diets without negative effects on

Additives		Findings	Reference
Probiotic (fermentation)	Canola meal (20%)	• Improved body weight gain, and feed conversion ratio • Increased the population of Lactobacillus spp. and decreasing the *E. coli* • Enhanced body weight gain, and feed conversion ratio	[84]
Probiotic and enzymes	Canola meal (20%)	• Higher nutrient digestibility • Increased in the relative weight of the bursa of Fabricius • and antibody titer against Newcastle disease • Improved antioxidant capacity, and gut health.	[46]
Enzymes	Canola meal (17.5%)	• Higher final body weight and improved FCR • Highest levels of aspartate aminotransferase (AST). • Improved health status.	[89]
S. cerevisiae	Olive cake (10%)	• Enhanced body weight gain, and feed conversion ratio. • The best European production efficiency index in broilers.	[90]
Citric acid	Olive cake (20%)	• Increased protein utilization. • Enhanced body weight gain, and increased feed intake	[91]
Probiotic (*A. awamori*)	Olive pulp (15%)	• Enhanced productive performance • Improved nutrient digestibilities.	[86]
Probiotic (fermentation)	Rapeseed meal (10%)	• Improved the production performance and maintain good health. • lower the intestinal pH and improve the intestinal barrier function	[92]
Enzymes	Cottonseed meal	• Improved performance (lowered the FCR and increased BWG) • Enhanced the digestibility of amino acids, and starch	[93]
Probiotic (fermentation)	Cottonseed meal	• Decreased abdominal fat and hepatic triglycerides • Improves growth performance, gut microbes, • Strengthening the immune system and reinforcing stress fighting capabilities	[94]
Enzymes	Sunflower meal	• Improved growth performance • Increasing nutrient digestibility	[95]

Table 3.
Using feed additives to improve the utilization of some unconventional feed ingredients in poultry feeding.

performance [98]. Many studies reported that sunflower meal can be used at higher levels with no negative effects on the utilization and growth performance of broiler chickens with the addition of enzymes [95]. Supplementation of exogenous enzymes in poultry diets can decrease their deleterious effects on high concentrations of fiber and stimulate fiber digestion. Supplementation of exogenous enzymes functions in the breakdown of NSPs and reduction of gut viscosity, thus improving nutrients digestibility and gut performance. In some studies, an improvement in growth performance (LBW and FCR) was observed in the birds fed SFM supplemented with exogenous enzymes compared to the control diet [100, 101] maybe as a result of the enhancement of other physiological and metabolic processes such as depolymerize complex NSPs and increasing nutrient digestibility that has prebiotic effects on health-promoting bacterial proliferation by releasing fermentable manno-, galacto-, xylo-, or gluco-oligomers during cracking of fiber [95]. In addition, it increases energy concentration and enhances nutrient utilization and absorption [100, 101]. From that, sunflower meal up to 25–30% can be used with some feed additives as a soybean meal substitute.

4.3 Cottonseed meal

Cottonseed meal is a potentially good source of protein (41–44%) and metabolizable energy as a substitute for soybean in poultry diets, the protein percentage varies based on the degree of dehulling before oil extraction. However, using cottonseed meal as an ingredient in poultry feeds presents numerous challenges and limitations, including high fiber content, as well as high presence of gossypol (a toxic polyphenolic pigment) and unbalanced amino acids such as low lysine content. Many reports have shown various ways to optimize the use of cottonseed meal in poultry feed such as proper processing, supplementation with lysine, other feed additives, or a particular combination of feed additives [102]. Gossypol inhibits the activity of pepsin and trypsin in the gut, thereby reducing the digestibility of protein [94, 103]. One of the factors that can impede protein utilization and nutrient digestibility is the presence of non-starch polysaccharides (NSP), which the bird cannot digest. It is well established that supplementing exogenous enzymes in poultry feed, such as β-glucanase, xylanase, and pectinase, leads to improved digestibility by potentially lower intestinal content viscosity [102]. Several studies confirmed that adding a mixture of enzymes (β-glucanase and xylanase) to the poultry diet that contained cottonseed meal led to an improvement in performance (lowered the FCR and increased BWG) through improving the digestibility of amino acids, and starch [94]. The improvement in nutrient digestion may be due to the role of enzyme supplementation (β-glucanase and xylanase) in the removal of nutrient encapsulation of cell walls existing in many dietary ingredients, which leads to nutrient release and digestibility of dietary nutrients [93, 100]. Nevertheless in a previous study, the fermentation process was found very effective in detoxifying free gossypol, resulting in the improvement of the nutritional value of cottonseed meal [94]. Microbial fermentation is currently considered one of the most effective ways improvement of the nutritional value of unconventional feed *via* reducing anti-nutritional factors such as free gossypol [94, 102], thus enhancing the growth and health of poultry. In addition, the metabolic activities of probiotics microbials during fermentation lead to produce enzymes, vitamins, oligosaccharides, organic acids, and some other compounds, which help in improving growth performance, as well as enhancement of the nutritional value of cottonseed meal [94, 102]. Furthermore, fermentation could improve intestinal digestive enzyme activity, bacterial ecology, and intestinal morphology in broilers [85, 102]. From this, we can recommend that the fermentation process increases the amount of cottonseed meal that can be substituted for soybeans meal.

4.4 Olive pulp

Olive pulp (OP) is one of the olive oil extraction products, especially since it is rich in fatty acids (oleic, linoleic, and linolenic), crude protein, calcium, and copper [104, 105], as well as some biologically active compounds (polyphenol) that have an antioxidant, anti-inflammatory, and antibacterial properties [93]. However, OP has low nutritional value due to its low energy and indigestible proteins, in addition to containing high fiber content and lignin [105, 106]. It was necessary to use some feed supplements in chicken diets containing these by-products (such as OP) to reduce anti-nutritional factors, thereby reducing their negative influence on chicken production performance.

Some studies reported that the inclusion of up to 5% of olive pulp in the diet of broiler chickens did not have a negative effect on productive performance [107]. However, the addition of 10–15% olive pulp to the broilers had a detrimental effect on the health and performance of the bird [100]. This may be due to the anti-nutritional factors in olive pulp, especially the high fiber content (NSPs), which negatively influence gut ecology and thus health in monogastric animals, beside its high content of lignin which hinders the digestibility coefficients of nutrients.

Several studies have proven that the use of some feed additives has an effective effect in improving the nutritional value of olive pulp in poultry feed such as exogenous enzymes, probiotics, organic acid, etc. [103, 105, 106]. Some previous studies results indicated that adding *A. awamori* as a probiotic enhanced the nutritional value of the OP in the diet, leading to an improved broiler growth performance [87]. This improvement might be due to the reduction in the anti-nutritional factor of the feed and the improvement in gut health, by the activation of several health-promoting bacteria, improving the intestinal epithelial cells structure, and selectively stimulating their growth and immune system [87], as well as stimulating the activity of some enzymes (active amylase, glucoamylase, and protease) in the digestive system [108], leading to an improvement in the metabolism of protein and carbohydrates. Likewise, several studies reported that the inclusion of citric acid in rabbits' diets enhanced nutrient digestibility [108]. Organic acids contribute to improving nutrient digestibility by reducing antinutrients and adjusting gut pH, which stimulates the activity of beneficial microbes, reduces the number of pathogenic microbes, and reduces inflammation, thus contributing to the availability of nutrients [3, 91]. In addition, citric acid has been reported to enhance the utilization of protein and some minerals [108]. As a previous study indicated, feeding broiler chickens a diet containing olive cake (20%) and citric acid (1 g/kg) resulted in a positive effect on growth performance compared to chickens that were fed 20% olive cake without additives or control [91]. Previous studies indicate that combining olive pulp with enzymes, probiotics, or organic acids led to improves the nutritional value of olive pulp, this supports the possibility of replacing part of the diet with olive pulp.

5. Conclusion

This chapter confirms that the use of feed additives plays an important role in poultry feeding as safe alternatives to antibiotics and as improvements to the bird's performance under conditions of heat stress. Beside their importance to improve and enhance the utilization of feed, especially nonconventional feed materials. Based on the scientific findings, which have been mentioned, the following important conclusions could be summarized as follows:

Using feed additives in the poultry diets can reduce the negative impact of heat stress on poultry *via* reduced heat production (fat and betaine), capable of reacting with free radicals (Vitamins A, E, and C) leading to reducing lipid peroxidation, improved antioxidant parameters (Vitamin E, C, and Zn and Se), increased antioxidant enzyme activity (plant extracts, e.g., oregano and lemon), a modulating the gut microflora (probiotic and plant extracts).

Antibiotic alternatives have analogous advantages to antibiotics to enhance the well-being and the production performance of broiler chickens without human health challenges. The major provided effects of alternative feed additives include enhance digestion, increase absorbability of nutrients, improved nutrient availability, antioxidant activity, immune-modulating, antimicrobial, improve intestinal health by enhancement of gut integrity, modulating the host gut microflora, and intestinal barrier function. Moreover, it increases body weight and carcass weight, enhances feed conversion ratio, and the gut health of broilers.

The use of nonconventional feed materials is necessary to reduce the costs of poultry feed, but some feed additives must be used to reduce the anti-nutrition factors that hinder their use in feeding poultry. This study confirmed that feed additives improved the nutritional value of many unconventional raw materials that have been already in poultry feeding under the Egyptian and marginal environmental conditions of Egypt.

Further studies must be carried out on the use of feed additives concerning their role in poultry thriftiness and the physiological responses of poultry stocks under heat-stress conditions. The recent approaches of biotechnology and its role in feed additives and the importance of food processing should also be considered.

Abbreviations

CM	canola meal
FCR	feed conversion ratio
GSH-Px	glutathione peroxidase
LBW	live body weight
NSPs	non-starch polysaccharides
OC	organic acids
OP	olive pulp
Se	selenium
SOD	superoxide dismutase
T4	thyroxin
T3	triiodothyronine
Zn	zinc

Author details

Ahmed El-Baz* and Raafat Khidr
Desert Research Center, Mataria, Cairo, Egypt

*Address all correspondence to: dm.a.baz@gmail.com

References

[1] Rahman MA, Parvin MS, Sarker RR, Islam MT. Effects of growth promoter and multivitamin-mineral premix supplementation on body weight gain in broiler chickens. Journal of the Bangladesh Agricultural University. 2012;**10**(2):245-248

[2] Perić L, Žikić D, Lukić M. Application of alternative growth promoters in broiler production. Biotechnology in Animal Husbandry. 2009;**25**(5-6-1):387-397

[3] Elbaz AM, Ibrahim NS, Shehata AM, Mohamed NG, Abdel-Moneim AME. Impact of multi-strain probiotic, citric acid, garlic powder or their combinations on performance, ileal histomorphometry, microbial enumeration and humoral immunity of broiler chickens. Tropical Animal Health and Production. 2021;**53**:1-10

[4] Morsy AS. Effect of zeolite (Clinoptilolite) as a salinity stress alleviator on semen quality and hemato-biochemical parameters of Montazah cocks under South Sinai conditions. Research Journal of Animal and Veterinary Sciences. 2018;**10**(2):9-17

[5] Abdel-Moneim AE, Elbaz AM, Khidr RE, Badri FB. Effect of in ovo inoculation of Bifidobacterium spp. on growth performance, thyroid activity, ileum histomorphometry and microbial enumeration of broilers. Probiotic Antimicrobial Proteins. 2020;**12**:873-882

[6] Abdel-Moneim AME, Shehata AM, Mohamed NG, Elbaz AM, Ibrahim NS. Synergistic effect of Spirulina platensis and selenium nanoparticles on growth performance, serum metabolites, immune responses, and antioxidant capacity of heat-stressed broiler chickens. Biological Trace Element Research. 2022;**200**(2):768-779

[7] Elbaz AM, Ashmawy ES, Salama AA, Abdel-Moneim AME, Badri FB, Thabet HA. Effects of garlic and lemon essential oils on performance, digestibility, plasma metabolite, and intestinal health in broilers under environmental heat stress. BMC Veterinary Research. 2022;**18**(1):1-12

[8] Abdel-Moneim AME, Shehata AM, Khidr RE, Paswan VK, Ibrahim NS, El-Ghoul AA, et al. Nutritional manipulation to combat heat stress in poultry–A comprehensive review. Journal of Thermal Biology. 2021;**98**:102915

[9] Elbaz AM, Ahmed AM, Abdel-Maksoud A, Badran AM, Abdel-Moneim AME. Potential ameliorative role of Spirulina platensis in powdered or extract forms against cyclic heat stress in broiler chickens. Environmental Science and Pollution Research. 2022;**29**(30):45578-45588

[10] Hakimul H, Subir S, Shariful I, Aminul I, Rezaul K, Mohammad EHK. Sustainable antibiotic-free broiler meat production: Current trends, challenges, and possibilities in a developing country perspective. Biology. 2020;**9**:1-24

[11] Mehdi Y, Létourneau-Montminy MP, Gaucher ML, Chorfi Y, Suresh G, Rouissi T, et al. Use of antibiotics in broiler production: Global impacts and alternatives. Animal Nutrition. 2018;**4**(2):170-178

[12] Amaechi N, Amaeze PN. Effect of dietary chloroacetic acid as antibiotic replacer on the gastrointestinal microflora and gut morphology of

weanling pigs. Research Opinions in Animal and Veterinary Sciences. 2012;**2**:494-498

[13] Jet SM, Florencia NS. Phytogenic feed additives as an alternative to antibiotic growth promoters in poultry nutrition. Advanced studies in the 21st century. Animal Nutrition. 2021;**8**:1-18

[14] Abdel-Moneim AME, Shehata AM, Alzahrani SO, Shafi ME, Mesalam NM, Taha AE, et al. The role of polyphenols in poultry nutrition. Journal of Animal Physiology and Animal Nutrition. 2020;**104**:1851-1855

[15] Abdel-Moneim A-ME, Selim DA, Basuony HA, Sabic EM, Saleh AA, Ebeid TA. Effect of dietary supplementation of Bacillus subtilis spores on growth performance, oxidative status and digestive enzyme activities in Japanese quail birds. Tropical Animal Health and Production. 2020;**52**:671-680

[16] Abd El-Hack ME, El-Saadony MT, Shafi ME, Qattan SY, Batiha GE, Khafaga AF, et al. Probiotics in poultry feed: A comprehensive review. Journal of Animal Physiology and Animal Nutrition. 2020;**104**:1835-1850

[17] Thornton PK, J.van de Steeg, A. Notenbaert, M. Herrero. The impacts of climate change on livestock and livestock systems in developing countries: A review of what we know and what we need to know. Agricultural Systems. 2009;**101**:113-127

[18] Daghir NJ. Nutritional strategies to reduce heat stress in broilers and broiler breeders. Lohmann Information. 2009;**1**:6-15

[19] DEFRA (Department for Environment, Food and Rural Affairs). Heat stress in poultry, solving the problem. UK. Nobel House 17 Smith Square London SW1P 3JR. 2005

[20] Attia Y, Abdalah A, Zeweil H, Bovera F, El-Din AT, Araft M. Effect of inorganic or organic selenium supplementation on productive performance egg quality and some physiological traits of dual-purpose breeding hens. Czech Journal of Animal Science. 2010;**55**:505-519

[21] Rao SR, Prakash B, Raju M, Panda A, Kumari R, Reddy EPK. Effect of supplementing organic forms of zinc, selenium and chromium on performance, anti-oxidant and immune responses in broiler chicken reared in tropical summer. Biological Trace Element Research. 2016;**172**:511-520

[22] Kucuk O. Zinc in a combination with magnesium helps reducing negative effects of heat stress in quails. Biological Trace Element Research. 2008;**123**:144-153

[23] Panda A, Ramarao S, Raju M, Chatterjee R. Effect of dietary supplementation with vitamins E and C on production performance, immune responses and antioxidant status of white Leghorn layers under tropical summer conditions. British Poultry Science. 2008;**49**:592-599

[24] Abudabos AM, Suliman GM, Al-Owaimer AN, Sulaiman ARA, Alharthi AS. Effects of nano emulsified vegetable oil and betaine on growth traits and meat characteristics of broiler chickens reared under cyclic heat stress. Animals. 2021;**11**(7):1911

[25] Attia YA, Abd El AE-HE, Abedalla AA, Berika MA, Al-Harthi MA, Kucuk O, et al. Laying performance, digestibility and plasma hormones in laying hens exposed to chronic heat stress as affected by betaine, vitamin C, and/or vitamin E supplementation. Springer Plus. 2016;5:1619

[26] Attia YA, Al-Harthi MA, ShElnaggar A. Productive, physiological and immunological responses of two broiler strains fed different dietary regimens and exposed to heat stress. Italian Journal of Animal Science. 2018;**17**:686-697

[27] Al-Harthi M, El-Deek A, Al-Harbi B. Interrelation ships among triiodothyronine (T3), energy and sex on nutritional and physiological responses of heat stressed broilers. Egyptian Poultry Science. 2002;**22**:349-385

[28] Habashy WS, Milfort MC, Fuller AL, Attia YA, Rekaya R, Aggrey SE. Effect of heat stress on protein utilization and nutrient transporters in meat-type chickens. International Journal of Biometeorology. 2017;**61**:2111-2118

[29] Babinszky L, Horváth M, Remenyik J, Verstegen MWA. The adverse effects of heat stress on the antioxidant status and performance of pigs and poultry and reducing these effects with nutritional tools. In: Hendriks WH, Verstegen MWA, Babinszky L, editors. Poultry and Pig Nutrition. Challenges of the 21st Century. Wageningen: Wageningen Academic Publishers; 2019. pp. 187-208

[30] Musharaf NA, Latshaw JD. Heat increment as affected by protein and amino acid nutrition. World's Poultry Science Journal. 1999;**55**:233-240

[31] Pursey KM, Davis C, Burrows TL. Nutritional aspects of food addiction. Current Addiction Reports. 2017;**4**:142-150

[32] Ghazalah A, Abd-Elsamee M, Ali A. Influence of dietary energy and poultry fat on the response of broiler chicks to heat therm. International Journal of Poultry Science. 2008;**7**:355-359

[33] El-Deep MH, Ijiri D, Ebeid TA, Ohtsuka A. Effects of dietary nanoselenium supplementation on growth performance, antioxidative status, and immunity in broiler chickens under thermoneutral and high ambient temperature conditions. The Journal of Poultry Science. 2016;**53**(4):274-283

[34] Habibian M, Ghazi S, Moeini MM. Effects of dietary selenium and vitamin E on growth performance, meat yield, and selenium content and lipid oxidation of breast meat of broilers reared under heat stress. Biological Trace Element Research. 2016;**169**:142-152

[35] Sahin K, Sahin N, Kucuk O, Hayirli A, Prasad A. Role of dietary zinc in heat-stressed poultry: A review. Poultry Science. 2009;**88**:2176-2183

[36] Lee SR. Cr itical role of zinc as either an antioxidant or a Prooxidant in cellular systems. Oxidative Medicine and Cellular Longevity. 2018;**2018**

[37] Saleh AA, Eltantawy MS, Gawish EM, Younis HH, Amber KA, Abd El-Moneim AE-ME, et al. Impact of dietary organic mineral supplementation on reproductive performance, egg quality characteristics, lipid oxidation, ovarian follicular development, and immune response in laying hens under high ambient temperature. Biological Trace Element Research. 2020;**195**:506-514

[38] Baxter MF, Greene ES, Kidd MT, Tellez-Isaias G, Orlowski S, Dridi S. Water amino acid-chelated trace mineral supplementation decreases circulating and intestinal HSP70 and proinflammatory cytokine gene expression in heat-stressed broiler chickens. Journal of Animal Science. 2020;**98**:skaa049

[39] Abidin Z, Khatoon A. Heat stress in poultry and the beneficial effects of ascorbic acid (vitamin C) supplementation during periods of heat

stress. World's Poultry Science Journal. 2013;**69**:135-152

[40] Sahin N, Onderci M, Sahin K, Gursu M, Smith M. Ascorbic acid and melatonin reduce heat-induced performance inhibition and oxidative stress in Japanese quails. British Poultry Science. 2004;**45**:116-122

[41] Daghir N. Broiler feeding and management in hot climates. Poultry Production in Hot Climate. 2008:227-260

[42] Sahin K, Sahin N, Kucuk O. Effects of chromium, and ascorbic acid supplementation on growth, carcass traits, serum metabolites, and antioxidant status of broiler chickens reared at a high ambient temperature (32°C). Nutrition Research. 2003;**23**:225-238

[43] Burkholder K, Thompson K, Einstein M, Applegate T, Patterson J. Influence of stressors on normal intestinal microbiota, intestinal morphology, and susceptibility to Salmonella enteritidis colonization in broilers. Poultry Science. 2008;**87**:1734-1741

[44] Varasteh S, Braber S, Akbari P, Garssen J, Fink-Gremmels J. Differences in susceptibility to heat stress along the chicken intestine and the protective effects of galacto-oligosaccharides. PLoS One. 2015;**10**(9):e0138975

[45] Abd El-Hack ME, Abdel-Moneim A-ME, Mesalam NM, Mahrose KM, Khafaga AF, Taha AE, et al. Nigella Sativa Seeds and its Derivatives in Poultry Feed, Black Cumin (Nigella Sativa) Seeds: Chemistry, Technology, Functionality, and Applications. Springer; 2021. pp. 265-296

[46] Elbaz AM, El-sheikh SE, Abdel-Maksoud A. Growth performance, nutrient digestibility, antioxidant state, ileal histomorphometry, and cecal ecology of broilers fed on fermented canola meal with and without exogenous enzymes. Tropical Animal Health and Production. 2023;**55**:46

[47] Ebeid T, Fathi M, Al-Homidan I, Ibrahim Z, Al-Sagan A. Effect of dietary probiotics and stocking density on carcass traits, meat quality, microbial populations and ileal histomorphology in broilers under hot-climate conditions. Animal Production Science. 2019;**59**:1711-1719

[48] Li W, Bai J, Li Y, Qin Y, Yu D. Effects of Bacillus subtilis on meat quality, nutrient digestibility and scrum biochemical index of broilers. Chinese Journal of Veterinary Science. 2014;**34**:1682-1685

[49] Saleh AA, Shukry M, Farrag F, Soliman MM, Abdel-Moneim A-ME. Effect of feeding wet feed or wet feed fermented by bacillus licheniformis on growth performance, histopathology and growth and lipid metabolism marker genes in broiler chickens. Animals. 2021;**11**:83

[50] Abd El-Hack ME, Abdelnour SA, Taha AE, Khafaga AF, Arif M, Ayasan T, et al. Herbs as thermoregulatory agents in poultry: An overview. Science of the Total Environment. 2020;**703**:134399

[51] Aljabeili HS, Barakat H, Abdel-Rahman HA. Chemical composition, antibacterial and antioxidant activities of thyme essential oil (Thymus vulgaris). Food and Nutrition Sciences. 2018;**9**:433

[52] Toghyani M, Tohidi M, Gheisari AA, Tabeidian SA. Performance, immunity, serum biochemical and hematological parameters in broiler chicks fed dietary thyme as alternative for an antibiotic growth promoter. African Journal of Biotechnology. 2010;**9**:6819-6825

[53] Sigolo S, Milis C, Dousti M, Jahandideh E, Jalali A, Mirzaei N, et al. Effects of different plant extracts at various dietary levels on growth performance, carcass traits, blood serum parameters, immune response and ileal microflora of Ross broiler chickens. Italian Journal of Animal Science. 2021;**20**:359-371

[54] Attia YA, Bakhashwain AA, Bertu NK. Thyme oil (thyme vulgaris L.) as a natural growth promoter for broiler chickens reared under hot climate. Italian Journal of Animal Science. 2017;**16**(2):275-282

[55] Khafar K, Mojtahedin A, Rastegar N, Kalvani Neytali M, Olfati A. Dietary inclusion of thyme essential oil alleviative effects of heat stress on growth performance and immune system of broiler chicks. Iranian Journal of Applied Animal Science. 2019;**9**:509-517

[56] Nazarizadeh H, Mohammad Hosseini S, Pourreza J. Effect of plant extracts derived from thyme and chamomile on the growth performance, gut morphology and immune system of broilers fed aflatoxin B1 and ochratoxin A contaminated diets. Italian Journal of Animal Science. 2019;**18**:1073-1081

[57] Alirezaei M, Gheisari HR, Ranjbar VR, Hajibemani A. Betaine: A promising antioxidant agent for enhancement of broiler meat quality. British Poultry Science. 2012;**53**:699-707

[58] He S, Zhao S, Dai S, Liu D, Bokhari SG. Effects of dietary betaine on growth performance, fat deposition and serum lipids in broilers subjected to chronic heat stress. Animal Science Journal. 2015;**86**:897-903

[59] Nasiroleslami M, Torki M, Saki AA, Abdolmohammadi AR. Effects of dietary guanidinoacetic acid and betaine supplementation on performance, blood biochemical parameters and antioxidant status of broilers subjected to cold stress. Journal of Applied Animal Research. 2018;**46**:1016-1022

[60] Liu W, Yuan Y, Sun C, Balasubramanian B, Zhao Z, An L. Effects of dietary betaine on growth performance, digestive function, carcass traits, and meat quality in indigenous yellow-feathered broilers under long-term heat stress. Animals. 2019;**9**:506

[61] Dos Santos TT, Dassi SC, Franco CR, Da Costa CR, Lee SA, Da Silva AVF. Influence of fibre and betaine on development of the gastrointestinal tract of broilers between hatch and 14 d of age. Animal nutrition. 2019;**5**:163-173

[62] United states Department of Agriculture. Restrictions on Antibiotic Use for Production Purposes in U.S. Livestock Industries Likely To Have Small Effects on Prices and Quantities. Economic Research Service. 2019

[63] Elghandour MMY, Tan ZL, Abu Hafsa SH, Adegbeye MJ, Greiner R, Ugbogu EA, et al. Saccharomyces cerevisiae as a probiotic feed additive to non and pseudo-ruminant feeding: A review. Journal of Applied Microbiology. 2020;**128**(3):658-674

[64] Zhou M, Tao Y, Lai C, Huang C, Zhou Y, Yong Q. Effects of mannanoligosaccharide supplementation on the growth performance, immunity, and oxidative status of partridge shank chickens. Animals. 2019;**9**(10):817

[65] Mahmoud H, Afiffy O, Mahrous M. Effect of using formic acid on growth performance and some blood parameter of broiler chicken. Assiut Veterinary Medical Journal. 2020;**66**(164):140-154

[66] Kaczmarek SA, Barri A, Hejdysz M, Rutkowski A. Effect of

different doses of coated butyric acid on growth performance and energy utilization in broilers. Poultry Science. 2016;**95**(4):851-859

[67] Abd El-Hack ME, Alaidaroos BA, Farsi RM, Abou-Kassem DE, El-Saadony MT, Saad AM, et al. Impacts of supplementing broiler diets with biological curcumin, zinc nanoparticles and bacillus licheniformis on growth, carcass traits, blood indices, meat quality and cecal microbial load. Animals. 2021;**11**(7):1878

[68] Al Fatah MA. Probiotic modes of action and its effect on biochemical parameters and growth performance in poultry. Iranian Journal of Applied Animal Science. 2020;**10**:9-15

[69] Celina EB, Tamiris NSS. Probiotics as a promising additive in broiler feed: Advances and limitations. In: Advances in Poultry Nutrition Research. IntechOpen; 2021

[70] Kaur AP, Bhardwaj S, Dhanjal DS, Nepovimova E, Cruz-Martins N, Kuča K, et al. Plant prebiotics and their role in the amelioration of diseases. Biomolecules. 2021;**11**(3):440

[71] Yaqoob M, Abd El-Hack ME, Hassan F, El-Saadony MT, Khafaga A, Batiha G, et al. The potential mechanistic insights and future implications for the effect of prebiotics on poultry performance, gut microbiome, and intestinal morphology. Poultry Science. 2021;**100**(7):101143

[72] Indikova I, Humphrey TJ, Hilbert F. Survival with a helping hand: campylobacter and microbiota. Frontiers in Microbiology. 2015;**6**:1-6

[73] Chee SH, Iji PA, Choct M, Mikkelsen LL, Kocher A. Characterization and response of intestinal microflora and mucins to manno-oligosaccharide and antibiotic supplementation in broiler chickens. British Poultry Science. 2010;**51**:368-380

[74] Partanen KH, Morz Z. Organic acids for performance enhancement in pig diets. Nutrition Research Reviews. 1999;**12**:117-145

[75] Christian L, Mellor S. The use of organic acids in animal nutrition, with special focus on dietary potassium deformity under European and Austral-Asian conditions. Recent Advances in Animal Nutrition – Australia. 2011;**4**:123-130

[76] Adil S, Banday MT, Bhat GA, Mir MS, Rehman M. Effect of dietary supplementation of organic acids on performance, intestinal histomorphology, and serum biochemistry of broiler chicken. Veterinary Medicine International. 2010;**8**:1-7

[77] Hassan HMA, Mohamed MA, Youssef AW, Hassan ER. Effect of using organic acids to substitute antibiotic growth promoters on performance and intestinal microflora of broilers. Asian Australasian Journal of Animal Sciences. 2010;**23**:1348-1353

[78] Chowdhury R, Islam KMS, Khan MJ, Karim MR, Haque MN, Khatun M, et al. Effect of citric acid, avilamycin, and their combination on the performance, tibia ash, and immune status of broilers. Poultry Science. 2009;**88**(8):1616-1622

[79] Madhupriya V, Shamsudeen P, Raj Manohar G, Senthilkumar S, Soundarapandiyan V, Moorthy M. Phyto feed additives in poultry nutrition - A review. International Journal of Science, Environment and Technology. 2018;**7**:815-822

[80] Li HL, Zhao PY, Lei Y, Hossain MM, Kim IH. Phytoncide, phytogenic feed

additive as an alternative to conventional antibiotics, improved growth performance and decreased excreta gas emission without adverse effect on meat quality in broiler chickens. Livestock Science. 2015;**181**:1-6

[81] Miguel MG. Antioxidant and anti-inflammatory activities of essential oils: A short review. Molecules. 2010;**15**:9252-9287

[82] Joshua JF, Bhaskar G, Sami D. Phytogenic feed additives improve broiler feed efficiency via modulation of intermediary lipid and protein metabolism–related signaling pathways. Poultry Science. 2021;**100**:1-11

[83] Elbaz AM, Salama AA, Ashmawy ES, Thabet HA, Badri FBA. Effect of dietary inclusion of corn distillers dried grains with solubles on productive performance, nutrient digestibility, immune status, Ileal Histometric and microbiota of broiler chickens. Egyptian Journal of Nutrition and Feeds. 2022;**25**(3):369-376

[84] Elbaz AM. Effects of the diet containing fermented canola meal on performance, blood parameters, and gut health of broiler chickens. Journal Worlds Poultry Research. 2021;**11**(1):1-7

[85] Abdel-Moneim A-ME, Sabic E, Abu-Taleb A, Ibrahim N. Growth performance, hemato-biochemical indices, thyroid activity, antioxidant status, and immune response of growing Japanese quail fed diet with full-fat canola seeds. Tropical Animal Health and Production. 2020;**52**:1853-1862

[86] Abd El-Moneim AE, Sabic EM. Beneficial effect of feeding olive pulp and aspergillus awamori on productive performance, egg quality, serum/yolk cholesterol and oxidative status in laying Japanese quails. Journal of Animal and Feed Sciences. 2019;**28**:52-61

[87] Abd El-Moneim AE, Sabic EM, Abu-Taleb AM. Influence of dietary supplementation of irradiated or non-irradiated olive pulp on biochemical profile, antioxidant status and immune response of Japanese quails. Biological Rhythm Research. 2019;**53**:519-534

[88] Alshelmani MI, Loh TC, Foo HL, Sazili AQ, Lau WH. Effect of feeding different levels of palm kernel cake fermented by Paenibacillus polymyxa ATCC 842 on nutrient digestibility, intestinal morphology, and gut microflora in broiler chickens. Animal Feed Science and Technology. 2016;**216**:216-224

[89] Disetlhe ARP, Marume U, Mlambo V. Humic acid and enzymes inclusion in canola-based diets generate different responses in growth performance, protein utilization dynamics, and hemato-biochemical parameters in broiler chickens. Poultry Science. 2018;**97**(8):2745-2753

[90] Al-Harthi MA. The efficacy of using olive cake as a by-product in broiler feeding with or without yeast. Italian Journal of Animal Science. 2016;**15**(3):512-520

[91] Al-Harthi MA, Attia YA. Effect of citric acid on the nutritive value of olive cake in broiler diets. European Poultry Science. 2016;**80**:1-14

[92] Wu Z, Chen J, Ahmed Pirzado S, Haile TH, Cai H, Liu G. The effect of fermented and raw rapeseed meal on the growth performance, immune status and intestinal morphology of broiler chickens. Journal of Animal Physiology and Animal Nutrition. 2022;**106**(2):296-307

[93] Saki AA, Pournia K, Tabatabaie MM, Zamani P, Haghighat M, Salary J. Effects of cottonseed meal supplemented with

lysine and enzyme (Hydroenzyme XP) on egg quality and performance of laying hens. Revista Brasileira de Zootecnia. 2012;**41**:2225-2231

[94] Nie CX, Zhang WJ, Wang YQ, Liu YF, Ge WX, Liu JC. Tissue lipid metabolism and hepatic metabolomic profiling in response to supplementation of fermented cottonseed meal in the diets of broiler chickens. Journal of Zhejiang University. Science. B. 2015;**16**:447-455

[95] Mbukwane MJ, Nkukwana TT, Plumstead PW, Snyman N. Sunflower meal inclusion rate and the effect of exogenous enzymes on growth performance of broiler chickens. Animals. 21 Jan 2022;**12**(3):253

[96] Jazi V, Boldaji F, Dastar B, Hashemi SR, Ashayerizadeh A. Effects of fermented cottonseed meal on the growth performance, gastrointestinal microflora population and small intestinal morphology in broiler chickens. British Poultry Science. 2017;**58**:402-408

[97] Chiou PWS, Chen C, Yu B. Effects of Aspergillus oryzae fermentation extract on in situ degradation of feedstuffs. Asian-Australasian Journal of Animal Sciences. 2000;**13**(8):1076-1083

[98] Rama Rao S, Raju M, Panda A, Reddy M. Sunflower seed meal as a substitute for soybean meal in commercial broiler chicken diets. British Poultry Science. 2006;**47**:592-598

[99] Nassiri Moghaddam H, Salari S, Arshami J, Golian A, Maleki M. Evaluation of the nutritional value of sunflower meal and its effect on performance, digestive enzyme activity, organ weight, and histological alterations of the intestinal villi of broiler chickens. Journal of Applied Poultry Research. 2012;**21**:293-304

[100] Horvatovic M, Glamocic D, Zikic D, Hadnadjev T. Performance and some intestinal functions of broilers fed diets with different inclusion levels of sunflower meal and supplemented or not with enzymes. Brazilian Journal of Poultry Science. 2015;**17**:25-30

[101] Tüzün AE, Olgun O, Yıldız AÖ, Sentürk ET. Effect of different dietary inclusion levels of sunflower meal and multi-enzyme supplementation on performance, meat yield, ileum Histomorphology, and pancreatic enzyme activities in growing quails. Animals. 2020;**10**:680

[102] Niu JL, Wei LQ, Luo YQ, Yang WT, Lu QC, Zheng XX, et al. Fermented cottonseed meal improves production performance and reduces fat deposition in broiler chickens. Animal Bioscience. 2021;**34**(4):680

[103] Zhang WJ, Xu ZR, Zhao SH, Sun JY, Yang X. Development of a microbial fermentation process for detoxification of gossypol in cottonseed meal. Animal Feed Science and Technology. 2007;**135**:176-186

[104] Elbaz AM, Thabet HA, Gad GG. Productive and physiological performance of broilers fed diets containing different levels of olive pulp. Journal of Animal and Poultry Production. 2020;**11**(11):435-439

[105] Al-Harthi MA. The chemical composition and nutrient profiles and energy values of olive cake for poultry diets. Life Science Journal. 2014;**11**(9):159-165

[106] Zarei M, Ehsani M, Torki M. Productive performance of laying hens fed wheat-based diets included olive pulp with or without a commercial enzyme product. African Journal of Biotechnology. 2011;**10**:4303-4312

[107] Ibrahim NS, Sabic EM, Abu-Taleb AM. Effect of inclusion irradiated olive pulp in laying quail diets on biological performance. Journal of Radiation Research and Applied Sciences. 2018;**11**(4):340-346

[108] Nourmohammadi R, Khosravinia H, Afzali N. Effects of high dietary levels of citric acid on productive performance, serum enzyme activity, calcium and phosphorus retention and immune response in broiler chickens. European Poultry Science. 2015;**79**:1-9

Chapter 4

Occurrence of Hyperhomocysteinemia in Broilers and Reduction of Its Harmful Effects with Betaine- and Berberine-Supplemented Diets

Judit Remenyik, Ildikó Noémi Kovács-Forgács, Georgina Pesti-Asbóth, Ferenc Gál, Orsolya Csötönyi, László Babinszky and Veronika Halas

Abstract

Homocysteine is a metabolic intermediate in the methionine-cysteine conversion. High level of homocysteine in blood leads to changes in methylation pathways and consequently in transcriptional activation; therefore, it can disrupt gene expression. This chapter presents the biochemical pathways of the transformation of homocysteine in broilers and demonstrates the beneficial effects of certain bioactive feed additives (betaine and berberine) to health-related and production problems caused by the accumulation of homocysteine. Based on recent scientific findings, the following conclusions have been drawn: Hyperhomocysteinosis has received little attention in the field of avian physiology research. Currently used feed additives, such as betaine, potentially decrease circulating homocysteine, but support only one of the pathways responsible for homocysteine decomposition. Various phytonutrients may be suitable owing to their pleiotropic bioactive components, such as berberine. It can potentially maintain redox homeostasis in animals and modulate immune responses and therefore may be able to provide for liver protective functions. Additionally, it can encourage healthy tissue to express enzymes that are responsible for the degradation of homocysteine. Further studies are recommended to investigate how effectively berberine can reduce the incidence of hyperhomocysteinemia in broilers and whether it is necessary to use feed supplements throughout the life cycles of birds.

Keywords: broiler, hyperhomocysteinemia, metabolic disease, performance, betaine, berberine

1. Introduction

Broiler performance and meat quality are influenced by several factors, such as genotype, chick quality, feed consumption and nutrient supply, water supply, vaccination and health status, as well as conditions associated with housing, including stocking density, temperature and ventilation, lighting, litter quality and biosecurity [1]. All these factors also have important roles to play in profitable broiler meat production. Of these, the health status of birds has recently come to the focus of growing professional attention.

According to the calculations of Oxford Analytica (2023), in 2018 the global poultry production dropped by 2.8 million tons due to diseases [2]. During the same time, global egg production decreased by 3 million tons as a result of diseases, adding up to a loss of $5.6 billion in revenues.

Diseases occurring in poultry farming can be categorized in several ways [3]. One way of categorization is when diseases are classified by their common, such as genetic, mechanical, toxic and nutritional, causes. In another form of categorization, diseases are grouped as infectious and non-infectious diseases. Infectious diseases are caused by bacteria, viruses or fungi, whereas parasitic diseases are induced by external parasites, such as protozoa, worms, mites and lice. Unlike infectious diseases, non-infectious diseases originate from non-pathogenic organisms, and consequently they cannot be transferred from one animal to another.

Non-infectious diseases are caused by factors, such as genetics, malnutrition, environment, housing, etc. This group of diseases encompasses numerous illnesses, with the most common metabolic diseases in poultry being avian gout, dysbacteriosis, cage layer fatigue, fatty liver syndrome (FLS), fatty liver and kidney syndrome (FLKS), toxic fat syndrome (chick edema disease), ascites (AS), sudden death syndrome and spiking mortality syndrome [4]. In the poultry industry, metabolic problems have intensified in the past few decades, as the genetic potential of poultry for growth and feed efficiency has rapidly improved. The associated data further underpin that metabolic diseases are particularly common in broilers [4, 5].

In general, metabolic disorders can be chronic, impacting a relatively small percentage of the flock, or acute, meaning they affect a larger proportion of birds, while their incidence is often sporadic [6]. Metabolic diseases occurring in the cardiovascular system are responsible for a significant part of mortality in poultry stocks. The metabolic illnesses of the musculoskeletal system account for lower mortality rates, still they generally lead to slower growth and lameness [4, 7].

It has been revealed in human patients that with vascular diseases, plasma homocysteine (Hcy) level tends to be much higher than normal [8]. When homocysteine concentration in the blood is over the normal value, the resulting condition is called hyperhomocysteinemia. According to a review by Jakubowski and Son and Lewis, hyperhomocysteinemia in humans has been associated with cardiovascular, cerebrovascular and thromboembolic conditions, as well as hip fracture, osteoporosis, chronic kidney disease (CDK), hypothyroidism and mental problems, such as cognitive decline and Alzheimer's disease (AD) [9, 10]. It should be noted that hyperhomocysteinemia is also known in broilers, but this disorder is less documented and explored than in humans.

The central role of the transsulfuration pathway (TSP) in the development of non-infectious metabolic diseases has been demonstrated by research in recent years [11]. While *in vivo* experiments have been mostly performed with rodents (mice, rats), genomic studies have shown that it works through a conserved pathway that is likely to be present in all vertebrates, consequently for poultry, too. The key enzymes (cystathionine-γ-lyase (CSE) [EC 4.4.1.1] and cystathionine-β-synthase (CBS) [EC 4.2.1.22]) have been identified not only in rodents, but also in pigs and humans. The processes catalyzed by enzymes involved in the degradation of homocysteine are the only pathway for endogenous hydrogen sulfide (H_2S) production. The absence of these processes leads to the emergence of abiotic stress. The accumulation of homocysteine reduces the SAM- (S-adenosylmethionine) dependent transmethylase activity, leading to hypo-methylation, deoxyribonucleic acid (DNA) methylation and histone acetylation. These processes can disrupt (suppress or activate) gene expression, and thus compromise basic metabolic functions. Increased plasma homocysteine levels are negatively correlated with the H_2S concentrations in cells or tissues. As a consequence, inflammatory processes are triggered, which can result in the inflammation of the muscles, as well as plaque formation in the cardiovascular system or the emergence of respiratory diseases. Even before the symptoms of metabolic diseases can be observed, metabolic changes may adversely affect the performance of birds (broilers) and the profitability of meat production.

To support the transsulfuration pathway, natural feed additives can be used for their potential to bring down pathologically high homocysteine levels in the plasma. However, these additives can be used efficiently in broiler nutrition only if the biochemical pathway of homocysteine formation is precisely known, and it is also clear how to modify the pathway of transformation of homocysteine into cysteine (Cys) or methionine (Met) with the use of the particular feed additives in the broilers [12].

By way of this review, we want to provide guidance to nutrition professionals in this intricate field of expertise by systemizing and evaluating the latest scientific findings. Therefore, the purpose of this chapter is to describe the biochemical pathways of the transformation of homocysteine into methionine or cysteine in broilers alongside the health-related and production problems that are caused by the large-scale accumulation of homocysteine. A further goal is to demonstrate the beneficial effects of some bioactive feed additives (betaine and berberine (BBR)) on the incidence of hyperhomocysteinemia and the performance of birds.

2. Systematic review methodology

The keywords used to research and collect the literature for this critical review included: "homocysteine," "transsulfuration pathway," "broiler," "methionine," "cysteine," "betaine" and "berberine," either individually or in a combination thereof. Databases searched included PubMed, ScienceDirect, Google Scholar, Scopus and Web of Science. The publication period beginning from 2001 was chosen as a starting point, with the recency of research as the prime focus for the inclusion of the majority of the studies. Subsequently, searching was enlarged from 1990 in order to study the wider perspective. Close to 200 journal articles satisfied the criteria, and after review, 109 were shortlisted for inclusion in this review.

3. Vital roles of methionine and cysteine in broiler nutrition

Homocysteine is a metabolic intermediate in the methionine-cysteine conversion. Therefore, prior to discussing the problem associated with hyperhomocysteinemia, it seems to be useful to overview the role of the sulfur-containing amino acids concerned in poultry feeding.

3.1 Methionine

An essential amino acid, methionine (Met) plays several important roles in bird metabolism, such as protein synthesis and feather development; furthermore, it serves as a methyl group and sulfur donor for methylation and transsulfuration reactions, respectively. Moreover, it is a precursor of some key intermediates (cysteine [Cys], carnitine, S-adenosylmethionine, glutathione (GSH), taurine, etc.) in metabolic pathways [13–15]. Methionine is the principal donor of methyl radical in the body [16].

This amino acid is also considered as the first limiting amino acid for optimal growth of poultry on corn-soy diets, and therefore has particular significance [14]. Methionine has an essential role in energy production and boosts the livability, performance and feed efficiency in poultry. Beyond its fundamental importance in protein synthesis, methionine also exerts functional roles through its antioxidant capacities [17, 18]. It is also well known that methionine supplementation improves immune response through its direct effects on protein synthesis and breakdown, as well as owing to its indirect effects on the various derivatives of methionine [17, 19]. Methionine deficiency directly and negatively impacts broiler production. In such cases, weight gain, feed efficiency and protein content in the carcass are all reduced. When methionine deficiency is not drastic, feed intake slightly increases, and it contributes to the generation of extra energy in body. However, it results in more massive accumulation of body fat [20].

Amino acids, including methionine and cysteine, are mainly absorbed through the small intestine. In association with the amino acid demand of broilers, the amino acid content of feed ingredients and compound feeds are expressed as digestible amino acids. The amino acid content of diets should be determined in relation to the lysine (Lys) content (100%) with reliance on the ideal protein concept for broilers. That ensures the precise supply of amino acids and the maximum efficiency of their utilization for the smallest possible metabolic load.

The methionine content of diets is usually supplemented with industrially produced methionine. The common sources of methionine in broiler diets are DL-Met (a racemic mixture of the D- and L-isomers of Met in equal proportions), the hydroxy analogue of methionine calcium salt and the hydroxy analogue of methionine to bring the sulfur-containing amino acid content of diets into equilibrium so as to meet the needs of birds [21, 22]. Methionine sources are used in two distinct forms: powder and liquid. The latter one is a DL-Methionine hydroxy analogue-free acid (MHA-FA, containing 88% of the active substance) [23]. In general, conventional methionine products contain 50% of both D- and L-Met, but due to the fact that the animal body can utilize only L-amino acids in protein synthesis, a specific methionine product containing only L-Met has been developed. According to the product specifications, the methionine produced by the industry contains 98.5% L-Met, 0.5% water (loss on drying) and 0.1% ash [17]. However, it seems that the pure L-Met is not that important in practical feeding. In fact, normal metabolism features an efficient process of conversion of D-Methionine into L-Methionine. This two-step reaction encompasses the action of amino acid oxidase to remove the amine group from D-Met, which results in the generation of α-keto-methionine, to which transaminase attaches an amine group to form L-Methionine in the second step [24].

The methionine demand of broilers is influenced by various factors, such as growing phase, the type of production, sex and breed [17]. In their outstanding review, Rehman et al. conclude that different levels of methionine in poultry diets have been reported by researchers, ranging from 0.3% to 1.2% during the initial period and 0.3% to 0.9% in the growth period of poultry. Results also showed that more edible meat yield could be obtained by supplementing Met + Cyst at the rate of 80% of the digestible lysine [25].

Based on the results of Rehman et al., it can be concluded that if DL-Met and L-Met are included in diets at standard levels, they are equally effective as sources of methionine for broilers [25].

A study by Çenesiz et al. demonstrated that with methionine-deficient diet, the addition of this amino acid significantly improved the growth performance and carcass yield of broilers. Similarly to Rehman et al., this study suggested that no significant differences in growth performance and carcass quality parameters could be anticipated when broiler diets were supplemented with DL-Met or L-Met [22, 25].

Another study by Macelline et al. indicated an optimum methionine-to-lysine ratio of 50.3, which was somewhat higher than standard recommendations [26].

3.2 Cysteine

Even though there are four common sulfur-containing amino acids (methionine, cysteine, homocysteine and taurine), only methionine and cysteine are incorporated into proteins [27].

Cysteine is a semi-essential sulfur-containing amino acid. If the methionine-to-cysteine ratio is imbalanced, it causes depression in the growth of birds [28]. Methionine can be converted irreversibly to cysteine by transsulfuration. Therefore, in the feed tables demands for these amino acids are usually added up as methionine and cysteine requirements [20].

Cysteine plays an important role in a number of physiological processes. In addition to methionine, cysteine can improve intestinal histomorphometric indices of broilers [29], leading to an increase in the absorption of nutrients. Cysteine can prevent oxidative damage [30].

Baker points out that consuming more L-isomer of Cysteine (L-Cysteine) than necessary triggers acute metabolic acidosis in chickens [31]. In another study, Baker suggests that even at much larger doses, none of the known amino acids produce the same degree of lethality as excess L-Cysteine [32].

Cysteine, like other amino acids, is primarily absorbed through the small intestine.

Even if the physiological concentration of cysteine is adequate, lots of cells cover at least 47% of their cysteine demand via the transsulfuration pathway [33].

The ideal amino acid ratio relative to Lys was calculated to be 75% Met + Cys on a true fecal digestible basis [34]. Nearly a similar value was estimated by Baker and Han.

In most cases, broiler feed needs to be supplemented with methionine to ensure an adequate supply of Met + Cys to birds [16].

Compound feeds for broilers are formulated to meet methionine + cystine demands based on the assumption that dietary methionine is converted into cysteine [35].

In the literature, values for the Met + Cys demands of broilers vary broadly.

This can be attributed to several reasons, including differences in breeds, diet compositions, circumstances of animal studies, as well as differences in the bioavailability of methionine products used in the study; also, in many cases to the small number of animals per treatment and consequently, the large standard deviation of the mean values of treatments, etc.

Goulart et al. recommended 0.873, 0.755, 0.748 and 0.661% of digestible methionine + cystine in the diet for the pre-initial, initial, growing and final phases, respectively [20]. On the other hand, Millecam et al. found that the optimal methionine + cysteine levels for broilers are 0.69, 0.66 and 0.62% in the starter, grower and finisher phases, respectively. At present, it appears that more extensive research is needed to clarify the Met + Cys requirement for broilers [36].

In the light of the foregoing, it can be concluded that the metabolism of Met and Cys is closely interrelated, and both amino acids play important roles in the protein metabolism of broilers.

4. Methionine, cysteine and Hcy metabolism

Homocysteine is a sulfur-containing amino acid, a metabolic intermediate in the Met-Cys conversion. Based on stoichiometry, its circulating concentration is regulated by two key pathways: remethylation and transsulfuration (**Figure 1**). There is a third pathway, because Hcy can be converted to homocysteine thiolactone (HTL), but that conversion is active only when Hcy concentration is high. The following section discusses these different pathways.

4.1 Homocysteine transformation in the remethylation pathway

In the methionine cycle, homocysteine is produced from methionine in two steps. Briefly, after methionine adenosylation to S-adenosyl-L-Methionine (SAM/AdoMet) methyltransferase takes over the methyl group (DNA, ribonucleic acid (RNA), protein, phospholipid) to different acceptor molecules yielding S-adenosylhomocysteine (SAH/AdoHyc) as a by-product. The S-adenosylhomocysteine hydrolase cuts off the adenosine part and forms homocysteine. On the other hand, homocysteine

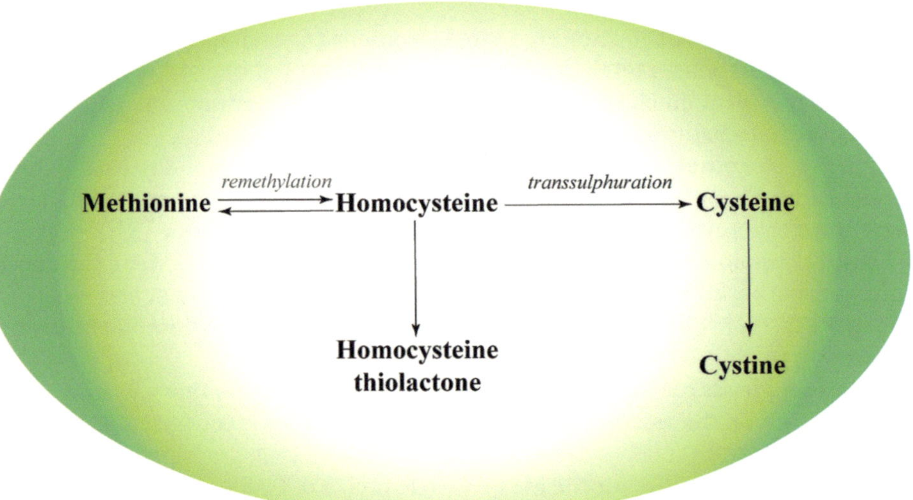

Figure 1.
Formation of homocysteine in the methionine-cystine pathway.

transforms into methionine with the methyl group from 5-*N*-methyl-tetrahydrofolate in a reaction catalyzed by vitamin B_{12}-dependent methionine synthase (**Figure 2**) [37]. As discussed by Vizzardi et al., high-level Hcy is accompanied by a reduced methylation potential, and therefore it compromises the Hcy → Met conversion, whereas folate and vitamin B_{12} tend to increase this potential. Although in the methyl group the main source of homocysteine remethylation into methionine is 5-*N*-methyl-tetrahydrofolate, betaine and choline also act as methyl donor molecules. The betaine pathway mostly occurs in the liver and is catalyzed by Hcy-methyltransferase [38].

4.2 Homocysteine transformation in the transsulfuration pathway

The transsulfuration pathway (TSP) accounts for the transformation of homocysteine into cysteine through cystathionine. It has a key role in sulfur metabolism and the redox environment of cells. TSP is the only way of cysteine biosynthesis in mammals and birds [39]. The first step is catalyzed by the vitamin B_6-dependent cystathionine-β-synthase (CBS) enzyme; homocysteine and serine are involved as substrates in the condensation reaction that produce cystathionine. The second step is a hydrolysis reaction that is catalyzed by the vitamin B_6-dependent cystathionine-γ-lyase (CSE). The substrate here is cystathionine, with cysteine and α-ketobutyrate (αKB) forming during the last step [37].

It should be noted, however, that in addition to remethylation and transsulfuration pathways homocysteine can undergo cyclization to form homocysteine thiolactone. This thioester is the toxic intermediate of homocysteine, as shown in **Figure 3**.

The role of the transsulfuration pathway in metabolic progresses is underlined by the fact that the essential H_2S signaling molecule is synthesized in this pathway. It has come in the focus of the scientific interests during recent years, because it has a principal physiological role, though inorganic H_2S smells like addled egg, and in

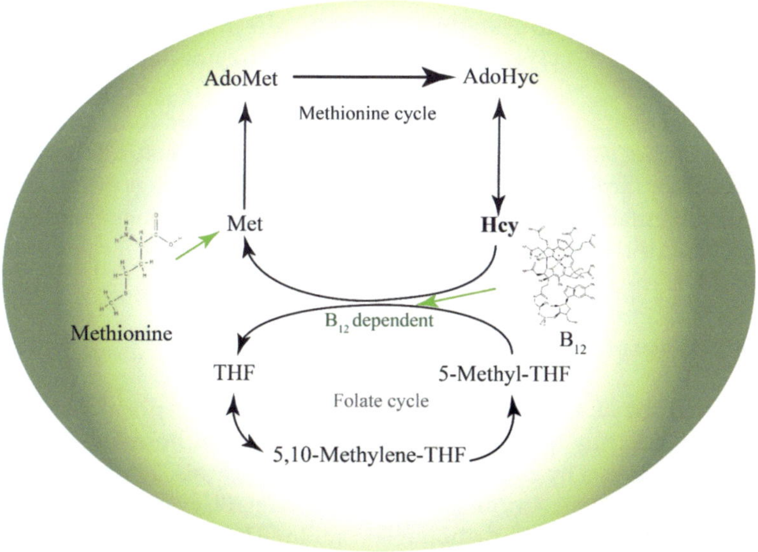

Figure 2.
Remethylation pathway of homocysteine and its dependency on the folate cycle. Abbreviations in the figure: tetrahydrofolate (THF); 5,10 methylene-tetrahydrofolate (5,10 methylene-THF); methionine (Met); S-adenosylmethionine (AdoMet); S-adenosylhomocysteine (AdoHcy); homocysteine (Hcy).

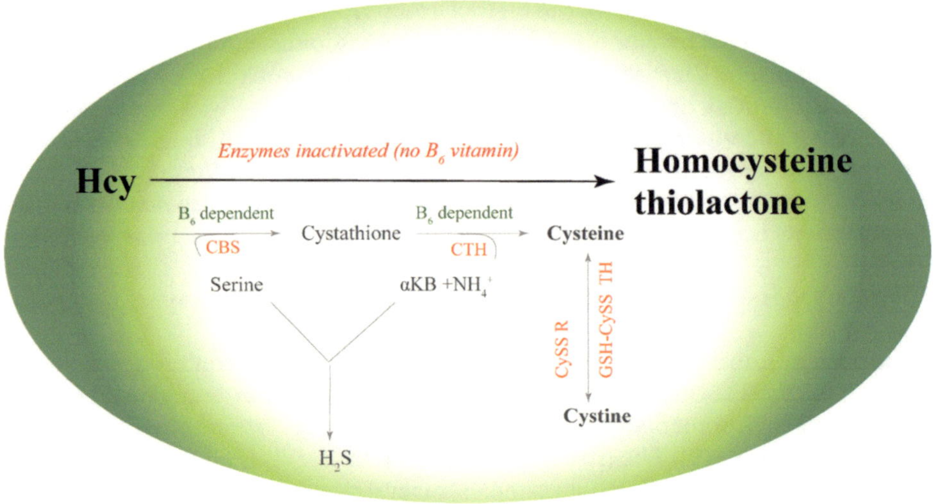

Figure 3.
Transsulfuration pathway and the formation of thiolactone in the absence or presence of vitamin B_6. Abbreviations in the figure: cystathionine-β-synthase (CBS); cystathione-γ-lyase (CSE); α-ketobutarate (αKB); ammonium ion (NH_4^+); cystine reductase (Cyss R); glutathione-cystine transhydrogenase (GSH-Cyss TH).

larger concentrations it is a toxic gas. The endogenous H_2S molecule was proven to be a vasoactive, cytoprotective, anti-inflammatory and antioxidant component. As a gas transmitter, it is able to diffuse through the cell membrane. Endogenous H_2S forms through enzymatic and non-enzymatic pathways in vertebrates. The former process is cytosolic and calls for mitochondrial enzymes: cystathionine-β-synthase (CBS), cystathionine-γ-lyase (CSE), 3-mercaptopyruvate sulfurtransferase (3-MST) and cysteine aminotransferase (CAT), by using L-Cysteine or homocysteine.

Hydrogen sulfide is produced by the enzymatic effect of CBS during the metabolism of Hcy-Cys disulfide into cystathionine. As an alternative way, H_2S also forms from cystine (disulfide of cysteine) as a result of the enzymatic effect of the CSE with ammonia (NH_3) release, while thiocysteine and pyruvate are produced. Thiocysteine splits into H_2S and Cys. Similarly to NO and carbon monoxide (CO), H_2S is a gaseous, fat-soluble messenger molecule. These three gas molecules constitute an unstable biological mediator family called gas transmitters. These findings point out that these molecules are enzymatically controlled and endogenously produced under normal physiological conditions in mammals, and therefore the biological roles of H_2S, NO and CO should be re-evaluated [37].

4.3 The importance of thiolactone

As mentioned above, the accumulated Hcy can easily transform into thiolactone, which is the reactive anhydride of homocysteine. Thioester homocysteine thiolactone (HTL) forms as a by-product of protein biosynthesis. The outcome of the process is that due to the structural similarities between Hcy and Met, during protein biosynthesis, methionyl-transfer ribonucleic acid (tRNA) synthetase builds into Hcy, instead of Met. Owing to repair mechanisms, homocysteine thiolactone is created. HTL forms isopeptide bonds with the residues of lysine (Lys). These isopeptide bonds lead to damaged or altered protein functions, and bring about pathophysiological effects, including autoimmune and intensified thrombosis activity. The HTL reaction with serum proteins induces the production of new protein antigens and autoimmune antibodies, which escalates inflammatory processes in the human and animal body.

Autoantibodies against the Nε-Hcy-Lys-protein complex can be found in the human plasma, and they positively correlate with the plasma total Hcy. Protein modulation mediated by HTL changes the protein sequence, which probably compromises the protein folding. Changes in the protein structure result in new interactions that influence cell physiology. Furthermore, HTL interacts with low-density lipoproteins (LDLs), which results in aggregation, increased density and vascular macrophage uptake, and also creates foam cells. Mammals, including humans, are able to eliminate the production of thiolactone through two distinct mechanisms. A high-density lipoprotein (HDL)-associated enzyme, Hcy-thiolactonase/paraoxonase-1 is capable of hydrolyzing Hcy-thiolactone both in the serum and intracellularly. Due to another mechanism, Hcy-thiolactone is decomposed by clearance in the kidney [9, 40].

In summary, homocysteine induces predisposition to metabolic dysfunctions, especially when its metabolite, homocysteine thiolactone, is formed at high levels. It has been reported in human studies that other N-homocysteinylated proteins also cause alteration in cell metabolism and trigger other mild or severe dysfunctions. N-homocysteinylated proteins can be cytotoxic and activate immune functions by forming immunoglobulin G (IgG) antibodies against N-homocysteinylated proteins to fight atherothrombosis [41, 42]. Homocysteine counteracts antioxidant enzymes and reduces their activity [43], and therefore hyperhomocysteinemia may deteriorate meat quality traits, too. The compromised antioxidant capacity predisposes to higher drip loss and may as well influence the color of meat.

5. Some metabolic diseases caused by high plasma homocysteine concentration (hyperhomocysteinemia)

5.1 Pathological conditions associated with hyperhomocysteinosis

The concentration of homocysteine, one of the cysteine metabolites, can potentially rise abnormally in the animal body, which may cause metabolic disorders; in severe cases, it may even lead to the death of the animal. In humans, the elevated concentration of circulating homocysteine may cause a number of cardiovascular diseases, such as heart attack, stroke, atherosclerosis and atherothrombosis [44, 45]. Therefore, the maintenance of plasma homocysteine balance in human patients can play an important role in the prevention of morbidity and mortality caused by cardiovascular diseases [46]. The normal level of Hcy in human adult plasma ranges from 5 to 15 μmol/L. In clinical routine, three hyperhomocysteinemia categories are distinguished: mild (15–30 μmol/L), moderate (30–100 μmol/L) and severe (>100 μmol/L). Hcy can be found in the circulation in its free form (approximately 1%), in disulfide, in a mixed form in disulfide and as bound to proteins, and therefore any assessment of real Hcy calls for the proper consideration of all these forms of Hys. The expression "total Hcy" (tHcy) is used for free Hcy, i.e., the quantity of Hcy from the reduction of disulfides together with the released, protein-bound Hcy from protein hydrolysis [47].

Despite the fact that elevated plasma homocysteine levels in humans have long been considered as a risk factor of cardiovascular diseases [44], this metabolic condition still has not received sufficient attention in poultry production. Unlike in the case of humans, the underlying reason is that in relation to livestock and poultry there is very limited information on the normal and pathological levels of Hcy. Nevertheless, it should be noted that in a few studies high levels of homocysteine have been revealed in specific metabolic disorders, such as sudden death syndrome, ascites, tibial dyschondroplasia and some myopathies that significantly compromise the profitability of poultry production. However, the number of poultry-related studies where plasma Hcy has been measured is extremely limited. For this reason, we consider it important to give a brief overview of some of the diseases that are caused by hyperhomocysteinemia and can also occur in the broiler industry.

5.2 Ascites and vascular diseases

In the chronic phase of ascites (AS), non-inflammatory transudate accumulates in one or more peritoneal cavities. The most common cause is the elevated hydraulic pressure originating right ventricular failure. In different phases, hepatic fibrosis is accompanied by these medical conditions. In the past, this health problem was noted only in the case of birds kept in high mountains, but nowadays it is a regular consequence of oxygen supply failure during rapid growth [48]. Wang et al. showed that broilers with cold-induced ascites suffered from severe liver failure, too [49].

In the course of AS, right ventricular failure triggers ventricular tachycardia and subsequently ventricular fibrillation, which are common consequences of coronary artery calcification. In atherosclerosis, vessel walls thicken, and fibrous caps emerge. If they turn into ulcers, endothelium gives rise to a coagulation cascade. The blood clot gets stuck in the vascular system or in the heart. The associated cause can be hyperlipidemia, which can induce liver failure and non-alcoholic fatty liver disease (NAFLD). Samuels found that the homocysteine concentration was three times higher in broilers with ascites than in healthy animals [50].

5.3 Skeletal disorders and myopathies

The tibial dyschondroplasia is a very common problem in intensively farmed, meat-producing poultry flocks. It is generally caused by the low mineralization of the tibia, the elevated calcium-phosphorus (Ca-P) ratio in the feed, fast primary osteon formation on the periosteal surface, as well as the insufficient filling of channels with osteoblasts [51]. Waqas et al. showed that during the development of the disease, alkaline phosphatase (ALP) and alanine aminotransferase (ALT) concentrations were rising in the plasma [52]. Orth et al. studied cysteine and homocysteine concentrations, and found that the homocysteine concentration of the plasma significantly rose in this musculoskeletal disorder [53].

Wooden breast syndrome (myopathy) is a systemic disease. It affects *pectoralis major* muscle on which pale, rib-like bulges and hemorrhages appear. Its development is associated with the abnormal accumulation of endomysial and perimysial connective tissues, with its consequences including fibrosis, hypoxia, oxidative stress and inflammatory responses to pathological conditions. There exists a well-known, close connection between wooden breast syndrome and hepatocyte injuries. In the plasma, aspartate aminotransferase (AST) and gamma-glutamyl transferase (GGT) concentrations rise with the elevated inflammatory cytokine profile. Maharjan et al. compared homocysteine concentrations in the plasma of healthy animals and animals with myopathy and found that Hcy concentration was very high in birds suffering from wooden breast syndrome [54]. Greene et al. also reported a 11-fold increase in the S-adenosylhomocysteine of breast meat categorized as wooden breast when compared to unaffected tissues [55].

6. Biochemical background of the homocysteine-lowering effects of some bioactive additives

6.1 Betaine

The consequences of hyperhomocysteinosis are disturbed methionine supply, low levels of available methionine on the cellular level and pathological conditions in response to high Hcy. Choline and betaine play a vital role in the remethylation of Hcy to Met. Choline is the parent compound of the class of cholines, consisting of ethanolamine residues with three methyl groups attached to the same nitrogen atom (**Figure 4**). It can be produced endogenously, but it is also often added in the form of choline chloride as a dietary supplement. Betaine is a generic name for a class of zwitterion compounds, but in nutritional science it is almost exclusively used to refer to glycine betaine or trimethyl glycine. The compound was first isolated from sugar beet, *Beta vulgaris*, hence the name. Betaine is the trimethyl derivative of glycine, the substrate of betaine-homocysteine S-methyltransferase (BHMT) in the liver and kidney. In the body, it can be found in anhydride, monohydrate, hydrochloride forms, which poultry can take advantage of. As an additive, the recommended concentration of betaine largely depends on the concentration of the methyl groups, environmental circumstances and the health status of birds. Similarly to choline, betaine is methyl donor in transmethylation processes, and consequently it can potentially decrease

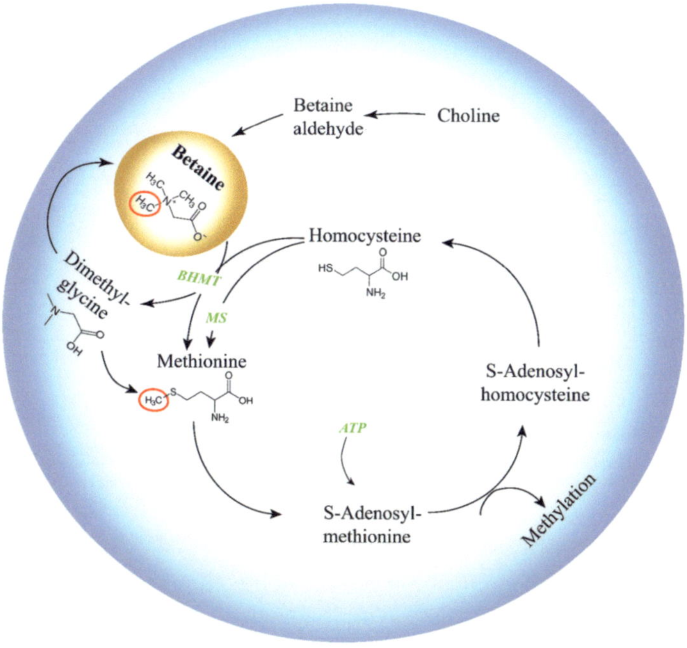

Figure 4.
Role of betaine in homocysteine transformation. Abbreviations in the figure: betaine-homocysteine S-methyltransferase (BHMT); methionine synthetase (MS), adenosine triphosphate (ATP).

creatine, methionine and choline demands. During the remethylation of homocysteine, betaine-homocysteine S-methyltransferase enzyme catalyzes the transfer of the methyl group from betaine into homocysteine, resulting in methionine and dimethyl-glycine. Therefore, for broilers dietary betaine is an alternative for methionine. Data presented in scientific papers are not clear-cut about the use of methionine in substitution for betaine. Studies also found that in the transsulfuration pathway the substitution of cysteine as an additive for betaine had more positive effects on the feed conversion ratio (FCR) in broilers than when given alone [56].

6.2 Berberine

Berberine is an isoquinoline alkaloid that can be found in a number of important herbs, such as *Berberis aristata* and *Berberis aquifolium*. Berberine features numerous pharmacological properties, including antibacterial, antihypertensive, anti-inflammatory, antidiabetic and liver protective effects [57–62].

It is evidenced in human patients that elevated plasma homocysteine levels are indicative of the increased risk of thrombotic and atherosclerotic vascular diseases [38] and steatosis, but berberine may be an effective substance to mitigate the risk of cardiac and metabolic diseases (**Figure 5**) [63]. It has been reported that hyperhomocysteinemia is associated with development of congestive heart failure in individuals who free from myocardial infarction. It induces systolic and diastolic dysfunction, arrhythmia, results in the accumulation of interstitial and perivascular collagen in the cardiac system and increases the risk of stroke [64–66]. In an experiment conducted with mice on homocysteine thiolactone-containing diet, the protective effects of berberine on the vascular function were revealed [67]. In another study,

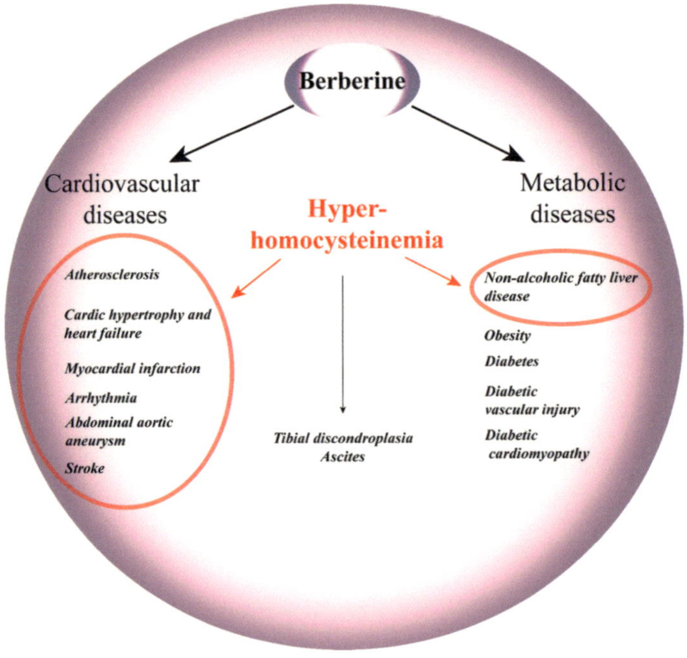

Figure 5.
Therapeutic potentials of berberine in different cardiometabolic diseases, some in association with hyperhomocysteinemia (adapted from Feng et al. [63]).

the area of atherosclerotic plaques could be reduced in mice that received berberine in the daily dose of 150 mg/kg BW per os [63]. Feng et al. have given a detailed overview of the mode of action by which berberine works in cardiac and other metabolic diseases [63].

One of the pathological conditions caused by hyperhomocysteinosis is the dysregulation of lipid metabolism and lipid accumulation in the liver, as a result of which the expression of CBS and CSE becomes damaged in liver tissues. The elimination pathways of homocysteine, such as remethylation and transsulfuration, are impaired in the course of hepatic steatosis. The biological activity of the natural phytochemical substance, berberine, has been studied in several animal experiments.

Homocysteine and cholesterol levels in the plasma of humans and animals with hyperhomocysteinemia have been found to be positively correlated [68–70]. The results of the relevant studies have confirmed a link between hyperhomocysteinemia and steatosis (fatty liver) [69–71]. Woo et al. found that homocysteine enhanced cholesterol secretion in the liver. The results suggested that hyperhomocysteinemia induced intensified cholesterol biosynthesis by regulating the corresponding transcriptomes; in fact, increased β-hydroxy-β-methylglutaryl-CoA (HMG-CoA) reductase gene expression was achieved in the liver [70]. In conclusion, both the liver and the serum cholesterol levels increase in hyperhomocysteinemia.

Wu et al. identified a mechanism by which berberine exerts a protective effect against cholesterol biosynthesis and liver dysfunction induced by homocysteine. Cholesterol synthesis was effectively limited by dietary berberine in rats with hyperhomocysteinemia. This inhibitory effect is mediated through the posttranslational modification of HMG-CoA reductase. Dietary berberine reduced cholesterol levels in the liver and improved liver function due to direct inhibition of HMG-CoA reductase (**Figure 5**) [11]. In line with the foregoing, Chang et al. also found in rats that berberine could counteract hyperhomocysteinemia and hyperlipidemia induced by high-fat diets, in part by upregulating low-density lipoprotein (LDL) receptor and apolipoprotein E (apoE) messenger ribonucleic acid (mRNA) levels, as well as by suppressing 3-hydroxy-3-methylglutaryl-CoA reductase gene expression. There is no direct correlation between HMG-CoA gene expression and changes in homocysteine levels. Although the results are not explanative of the direct cause of the decreasing homocysteine concentration in the plasma, experiments suggest that decreased lipid levels are associated with intact liver tissues and result in normal expression profiles [72].

Berberine is capable of decreasing oxidative stress; by supplementing feeds with berberine, the level of oxidative stress markers, such as malondialdehyde (MDA), glutathione (GSH), superoxide dismutase (SOD), becomes altered. Berberine inhibits reactive oxygen species (ROS) formation, improves mitochondrial function to boost the membrane potential and protects against oxidative damage. It moderates the activity of biomarkers mentioned above and increases the activity of antioxidant enzymes that help to bind free radicals and decrease oxidative stress [73]. Therefore, it may at least partly compensate for the compromised antioxidative defense mechanisms induced by hyperhomocysteinemia.

7. Some dietary means to reduce plasma homocysteine levels in broilers

As mentioned above, heart failure, i.e., one of the most frequently reported human diseases in the context of hyperhomocysteinemia, also tends to be a weighty problem in poultry production. Sudden death syndrome occurs typically with high-producing broilers and turkeys at the end of the fattening phase, shortly before slaughtering, and is accompanied by elevated plasma Hcy [50].

Several papers adopt the hypothesis of morbidity being the potential consequence of increased plasma homocysteine, but there are very few publications reporting measured homocysteine concentrations in birds. In a study with ducks, Xie et al. found increased plasma homocysteine concentration accompanied by decreased feed intake and compromised average daily gain (ADG) in response to an increase in dietary DL-Methionine from 0.285% to 0.685%. In line with that the foregoing, available data for broilers suggest that dietary methionine levels correlate with homocysteine levels in the plasma (**Figure 6**), while the oversupply of methionine results in elevated homocysteine levels in the blood [74]. According to Orth et al., tibial dyschondroplasia was observed to be more severe and frequent, with growth performance remaining poor when the diet was supplemented with homocysteine [53]. Authors, however, noted that bone deformation is probably not due to homocysteine, but may be attributed to the metabolite of homocysteine.

The metabolism of homocysteine requires B vitamins, particularly riboflavin, pyridoxal 5′-phosphate, cobalamin and folate. According to Lu et al., the primary pathway to maintain Hcy levels in the body is Hcy remethylation (61%), still a significant quantity of Hcy is catabolized into cysteine via transsulfuration (39%). The conversion of Hcy into cysteine is supported by betaine [75]. The results obtained by Ganson et al. also indicated that the folate-dependent remethylation of Hcy predominated over betaine-dependent remethylation, whereas betaine-dependent remethylation seemed to be more extensively influenced by dietary sulfur-containing amino acids [76]. Samuels confirmed that plasma Hcy levels decreased when broilers received diets supplemented with mixtures of pyridoxal, cobalamin, folic acid and betaine. Although there was 18% reduction in mortality from ascites and sudden death syndrome in the supplemented group, the difference was statistically not confirmed [50].

There are numerous studies shedding light on the effects of dietary trimethyl glycine (called betaine) on the performance of broilers. Most of these studies come to the conclusion that betaine is beneficial in the case of heat stress due to its role as a methyl donor and function as an intestinal and metabolic osmolyte [77]. It is known as an osmoregulatory substance that controls intracellular biochemical events, thus playing a key role in water balance during heat stress. Since it features three methyl groups, it serves as a methyl donor and can substitute choline or methionine in that

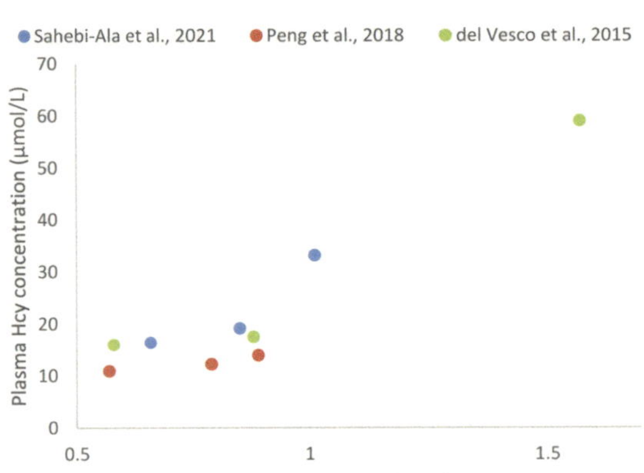

Figure 6.
Effect of dietary methionine plus cystine supply on plasma Hcy levels in broilers.

particular conversion. Sahebi-Ala et al. confirmed that during heat stress it was advisable to replace Met supplementation at least partly with betaine. In that study, 30% of the supplemental Met was replaced with betaine, which resulted in lower plasma homocysteine concentration [78]. Earlier studies also stressed the methionine-sparing effects of betaine [79–81].

In an outstanding review on the nutritional role of betaine, Abd El-Ghany and Babazadeh described a broad range of betaine supplementation that could be efficient in broiler chickens. Although in the cited literature, the dosage of supplementation ranged from 0.05 to 4 g/kg, with the most frequently applied levels of supplementation falling into the 1–2 g/kg range [82]. However, there are a very few studies performed with a focus on how betaine supplementation impacts Hcy levels. **Table 1** summarizes the broiler studies where plasma homocysteine levels were measured. These data demonstrate consistent reduction in plasma homocysteine concentrations in response to betaine supplementation, and the impact tends to be stronger in times of heat stress. The results presented by Mostashari-Mohases et al. show evidence that the regulation of plasma Hcy can be supported through the transsulfuration pathway, since betaine supplementation intensified betaine-homocysteine S-methyltransferase gene expression. In this broiler study, betaine supplementation has improved growth performance and feed efficiency, particularly in the finisher phase [83]. Recently, Maidin et al. have found that betaine decreased plasma homocysteine concentrations in the blood and improved the bone strength in laying hens [84]. Numerous publications have reported positive effects of betaine supplementation on growth performance in intensively farmed broilers [82, 85–87] and in slow-growing or indigenous broilers, too [88–90]. It has been revealed that through its role in methyl supply, betaine is able to support synthesis and increase the activities of enzymes that are responsible for antioxidant defense (glutathione peroxidase (GSH-Px), superoxide dismutase and glutathione). Despite its key role in metabolism, dietary betaine supplementation does not invariably result in better growth performance. There are a few studies that have not been able to confirm the improvement of body weight (BW) or growth rate with supplemental betaine, either in heat stress or in thermoneutral conditions [77, 78, 91, 92]. The ability of betaine to provide for the methyl group is beneficial in the Met→Cys conversion by supporting both the remethylation of Hcy to Met and the transsulfuration of Hcy to Cys. In this context, it plays a key role in maintaining Hcy levels and the reduction in the incidence of hyperhomocysteinemia in poultry. However, as mentioned above, Hcy levels have not been measured in most of the studies.

The metabolic load resulting from the limited ability to decompose homocysteine is likely to appear at later ages, and therefore—due to the short life cycle of broilers—growth performance may not be compromised on the flock level. Nevertheless, hyperhomocysteinemia is a metabolic challenge that predisposes to the sudden death syndrome and bone failure, as discussed above, and consequently can potentially decrease the economic efficiency of poultry farming.

According to relevant literature, among potential feed additives to reduce the incidence of hyperhomocysteinemia, berberine is one of the most promising candidates. While no direct evidence has been obtained for poultry, rat studies have confirmed that dietary berberine supplementation results in lower Hcy [11, 72]. Recently, berberine has come into the focus of interest for nutrition scientists, and numerous studies have been published discussing the health benefits related to its potential to mitigate oxidative stress, its anti-inflammatory and hepatoprotective potentials, as well as antimicrobial and antiviral activities. An excellent review by Imanshahidi and Hosseinzadeh pointed out that berberine exhibited multispectrum pharmacological

Reference	Feeding phase	Treatment	BW (g)	ADG (g/d)	ADFI (g/d)	CumFI (g)	FCR (g/g)	Mortality (%)	Effect	Plasma Hcy
Mostashari-Mohases et al. [83]	Phase 1	No betaine	1122			1711	1.59		Relative gene expression was 28-fold higher in the betaine-supplemented group	
		2 g/kg betaine suppl.	1180			1693	1.53			
	Phase 2	No betaine	2170^{a}			2254	1.88^{a}			
		2 g/kg betaine suppl.	2295^{b}			2240	1.73^{b}			
Kettunen et al. [77]	W0–3, ♀ (in heat stress)	Basal diet	Reported (but not confirmed) that BW was unaffected by betaine suppl.						Betaine supplementation reduced plasma Hcy	40.1 ± 2.5^{a} nmol/g
		1 g/kg betaine suppl.								30.6 ± 1.4^{b} nmol/g
	W0–3, ♂ (in heat stress)	Basal diet								47.8 ± 4.0^{a} nmol/g
		1 g/kg betaine suppl.								30.7 ± 1.5^{b} nmol/g
Samuels [50]	Phase 1 (W1–3)	Basal diet	892.8				1.328		Betaine supplementation reduced plasma Hcy	
		Supplemented diet[1]	875.1				1.306			
	Phase 2 (W4–6)	Basal diet	2723.8				2.02	9.7		35.9 μM/L
		Supplemented diet[1]	2666.6				2.01	8		29.7 μM/L
Maidin et al. [84]		No betaine	Tibia breaking strength and tibia density was improved by betaine						Betaine supplementation reduced plasma Hcy	20.3^{a} μM/L
		1 g/kg betaine suppl.								19.9^{b} μM/L

Reference	Feeding phase	Treatment	BW (g)	ADG (g/d)	ADFI (g/d)	CumFI (g)	FCR (g/g)	Mortality (%)	Effect	Plasma Hcy
Sahebi-Ala et al. [78]	Phase 1 (in heat stress)	Basal diet		53.57	78.56		1.487	1.22	Betaine supplementation reduced plasma Hcy	
		Supplemented diet[2]		54.48	78.77		1.463	1.70		
	Phase 2 (in heat stress)	Basal diet		75.54	151.77		2.056	3.09		23.49^{a} µmol/L
		Supplemented diet[2]		75.15	151.63		2.035	1.53		22.26^{b} µmol/L

BW, body weight; ADG, average daily gain; CumFI, cumulative feed intake; FCR, feed conversion ratio (gain/feed); Hcy, homocysteine.[1]Supplementation was 1 g/kg betaine, vitamin B_6 and B_{12}, as well as folate.
[2]Betaine supplementation was applied as substitution of 30% of methionine.
Differing superscripts in the same study indicate the differences of treatments ($P < 0.05$).

Table 1.
Effects of betaine supplementation on growth performance and plasma homocysteine concentrations in broiler studies.

action, ranging from cardiovascular conditions through anticancer effects to the modulation of antioxidants, neurotransmitters, enzymes, molecular targets and immune substances [93]. Moreover, berberine has been reported to influence energy, glucose and lipid metabolism [94]. These biological effects have made berberine an attractive natural compound that can be employed in sciences related to human and animal health [95].

As discussed above, homocysteine induces predisposition to metabolic dysfunctions. Homocysteine thiolactone results in protein damage, the aggregation and inactivation of functional proteins, such as enzymes and immune cells, and consequently enhances cell apoptosis [96]. Due to its property to support antioxidant defense mechanisms (**Table 2**), berberine may directly counteract Hcy and probably homocysteine thiolactone formation, or at least mitigate their negative impacts. Homocysteine counteracts antioxidant enzymes and reduces their activities [43], which is why hyperhomocysteinemia can potentially deteriorate the meat quality traits, too. It has been shown in broiler studies that the activities of antioxidant enzymes [97] and the characteristics of meat quality, particularly the water-holding capacity of meat in relation to oxidative damage to cells as induced by mycotoxin exposure, could be improved if the feed was supplemented with berberine [98, 99].

It has been reported in human studies that other N-homocysteinylated proteins also cause alteration in cell metabolism and result in various mild or severe dysfunctions. N-homocysteinylated proteins can be cytotoxic and activate immune functions by forming IgG antibodies against anti-N-homocysteinylated proteins [41, 42]. Berberine is a potential candidate to alleviate inflammatory mechanisms. Studies with broilers have confirmed reduction in ileal pro-inflammatory cytokines (interleukin-1beta (IL-1β), interleukin-6 (IL-6) and tumor necrosis factor alpha (TNF-α)) and lower intestinal necrosis indices [100, 101]. Fernandez et al. reported lower interleukin-17A (IL-17A), interleukin-17F (IL-17F), IL-6 and IL-1β in duck liver and spleen, when feed was supplemented with 200 mg/kg berberine [102]. As mentioned

Animal model	Dose of BBR	Treatment period (weeks)	Specimen used	Highlighted findings
Broilers	100 mg/kg/d	6	Serum	TAC↑, SOD↑, GSH-Px↑, MDA↓
Broilers	200, 400 and 600 mg/kg	6	Serum	SOD↑, GSH-Px↑, MDA↓
Broilers	200, 400 and 600 mg/kg	6	Meat	SOD↑, GSH-Px↑, MDA↓
Mice	200 mg/kg/d	2	Liver	SOD↑
Sprague-Dawley rats	80, 120 and 160 mg/kg/d	7	Serum	SOD↑
Sprague-Dawley rats	100 and 200 mg/kg/d	8	Kidney	SOD↑, MDA↓
Wistar rats	200 mg/kg/d	12	Serum	SOD↑, MDA↓
Wistar rats	75, 150 and 300 mg/kg/d	16	Serum and liver	SOD↑, GSH-Px↑, MDA↓

Abbreviations: TAC, total antioxidant capacity; GSH-Px, glutathione peroxidase; MDA, malondialdehyde; SOD, superoxide dismutase.

Table 2.
Effects of dietary berberine (BBR) supplementation on antioxidant defense in broilers and laboratory rodents (reviewed by Ghavipanje et al. [95]).

before, transmethylation and transsulfuration are parts of the basic metabolism and are known as conservative pathways. Consequently, we are convinced that it is worth using a trans-species approach to identify the effective feed additive that can reduce the emergence of hyperhomocysteinemia. To this end, in addition to betaine supplementation, berberine supplementation is one of the most potential candidates.

8. Conclusions

The following main conclusions can be drawn from the latest research findings:

- In poultry industry, metabolic diseases have intensified in the past few decades, as the genetic potential of poultry for growth and feed efficiency has improved rapidly.

- Even before the development of symptoms of metabolic diseases, metabolic changes may adversely affect the performance of birds (broilers) and the profitability of meat production.

- Homocysteine is a metabolic intermediate in the methionine-cysteine conversion. When homocysteine concentration in the blood is higher than normal, a condition called hyperhomocysteinemia occurs.

- Hyperhomocysteinosis is a well-known disorder in humans, but in the field of avian physiology it has received little attention in research. It may be useful to introduce the monitoring of homocysteine in poultry in order to understand and reveal the role of this metabolite in a number of systemic diseases.

- Currently used feed additives, such as betaine, are able to decrease plasma homocysteine concentrations, but they support only one of the pathways (transmethylation) responsible for homocysteine decomposition. However, other feed additives may as well be applied to activate the transsulfuration pathway, too. Various phytonutrients may be suitable owing to their pleiotropic bioactive components, such as berberine. This latter phytogenic feed additive may be capable of maintaining the redox homeostasis in animals by typically modulating inflammatory immune responses and may therefore be able to provide for liver protective functions. Furthermore, it can potentially encourage healthy tissues to express enzymes that are responsible for the degradation of homocysteine, such as cystathionine-β-synthase and cystathionine-γ-lyase.

- However, further studies are recommended to investigate how effectively berberine can reduce the incidence of hyperhomocysteinemia in broilers, and whether it is necessary to use feed supplements during the entire life cycle of the birds.

9. Conclusion for practice

In the case of fast-growing broilers, the prevalence of metabolic diseases may be on the rise. One of these frequently occurring metabolic diseases is hyperhomocysteinemia that can cause various conditions, such as dyschondroplasia, ascites and sudden death syndrome. In order to recognize these conditions on time, it is important to check the blood plasma homocysteine content of the flock. If the homocysteine content is higher than normal, it is advisable to add betaine to diets.

The concentration of betaine strongly depends on feeding, housing and health conditions. Therefore, it is recommended for farmers to conduct preliminary assessments in relation to the circumstances prevailing at the farms in question in order to determine how much betaine should be added to diets.

Abbreviations

3-MST	3-mercaptopyruvate sulfurtransferase
ADG	average daily gain
αKB	α-ketobutyrate
ALP	alkaline phosphatase
ALT	alanine aminotransferase
apoE	apolipoprotein E
AS	ascites
AST	aspartate aminotransferase
ATP	adenosine triphosphate
BBR	berberine
BHMT	betaine-homocysteine S-methyltransferase
BW	body weight
Ca	calcium
CAT	cysteine aminotransferase
CBS	cystathionine-β-synthase
CO	carbon monoxide
CSE	cystathionine-γ-lyase
CumFI	cumulative feed intake
Cys	cysteine
Cyss R	cystine reductase
DL-Met	D- and L-isomers of Met
DNA	deoxyribonucleic acid
FCR	feed conversion ratio
FLKS	fatty liver and kidney syndrome
FLS	fatty liver syndrome
GGT	gamma-glutamyl transferase
GSH	glutathione
GSH-Cyss TH	glutathione-cystine transhydrogenase
GSH-Px	glutathione peroxidase
H_2S	hydrogen sulfide
Hcy	homocysteine
HMG-CoA	β-hydroxy-β-methylglutaryl-coenzyme A
HTL	homocysteine thiolactone
IL-17A	interleukin-17A
IL-17F	interleukin-17F
IL-1β	interleukin-1beta
IL-6	interleukin-6
L-cysteine	L-isomer of cysteine
LDL	low-density lipoprotein

Lys	lysine
MDA	malondialdehyde
Met	methionine
MHA-FA	DL-methionine hydroxy analogue-free acid
mRNA	messenger ribonucleic acid
MS	methionine synthetase
NH_4^+	ammonium ion
NO	nitrogen-monoxide/nitric oxide
P	phosphorus
RNA	ribonucleic acid
ROS	reactive oxygen species
SAH/AdoHyc	S-adenosylhomocysteine
SAM	S-adenosylmethionine
SAM/AdoMet	S-adenosyl-L-methionine
SOD	superoxide dismutase
TAC	total antioxidant capacity
tHcy	total homocysteine
TNF-α	tumor necrosis factor alpha
tRNA	transfer ribonucleic acid
TSP	transsulfuration pathway

Author details

Judit Remenyik[1*], Ildikó Noémi Kovács-Forgács[1], Georgina Pesti-Asbóth[1], Ferenc Gál[1], Orsolya Csötönyi[2], László Babinszky[2,3] and Veronika Halas[2]

1 Center for Complex Systems and Microbiome Innovations, Faculty of Agricultural and Food Sciences and Environmental Management, University of Debrecen, Debrecen, Hungary

2 Department of Farm Animal Nutrition, Institute of Physiology and Nutrition, Hungarian University of Agriculture and Life Science, Kaposvár, Hungary

3 Faculty of Agricultural and Food Sciences and Environmental Management, Department of Animal Nutrition Physiology, University of Debrecen, Debrecen, Hungary

*Address all correspondence to: remenyik@agr.unideb.hu

References

[1] ArborAcres-Broiler Management Handbook. 2018. 1118-AVNAA-041. Available from: https://aviagen.com/assets/Tech_Center/AA_Broiler/AA-BroilerHandbook2018-EN.pdf [Accessed: February 4, 2024]

[2] Analytica O. Animal Health and Sustainability: A Global Data Analysis. Oxford Analytica; 2023. Available from: https://www.oxan.com/insights/global-data-healthforanimals/ [Accessed: May 31, 2024]

[3] Module 3: Common Poultry Diseases and Prevention Methods. Indiana State Poultry Association; n.d. https://www.inpoultry.com/module-3-national-poultry-improvement-plan-rules-and-forms [Accessed: February 2, 2024]

[4] Kumari A, Tripathi UK, Boro P, Sulabh S, Kumar M, Nimmanapalli R. Metabolic disease of broiler birds and its management: A review. International Journal of Veterinary Sciences and Animal Husbandry. 2016;**1**(3):15-16

[5] Angel R. Metabolic disorders: Limitations to growth of and mineral deposition into the broiler skeleton after hatch and potential implications for leg problems. Journal of Applied Poultry Research. 2007;**16**:138-149. DOI: 10.1093/japr/16.1.138

[6] Leeson S. Metabolic challenges: Past, present, and future. Journal of Applied Poultry Research. 2007;**16**:121-125. DOI: 10.1093/japr/16.1.121

[7] Julian RJ. Production and growth related disorders and other metabolic diseases of poultry—A review. Veterinary Journal. 2005;**169**:350-369. DOI: 10.1016/j.tvjl.2004.04.015

[8] Kaplan P, Tatarkova Z, Sivonova MK, Racay P, Lehotsky J. Homocysteine and mitochondria in cardiovascular and cerebrovascular systems. International Journal of Molecular Sciences. 2020;**21**:7698. DOI: 10.3390/ijms21207698

[9] Jakubowski H, Głowacki R. Chemical biology of homocysteine thiolactone and related metabolites. Advances in Clinical Chemistry. 2011;**55**:81-103. DOI: 10.1016/b978-0-12-387042-1.00005-8

[10] Son P, Lewis L. Hyperhomocysteinemia. Treasure Island (FL): StatPearls Publishing; 2024

[11] Wu N, Sarna LK, Siow YL, Karmin O. Regulation of hepatic cholesterol biosynthesis by berberine during hyperhomocysteinemia. American Journal of Physiology Regulatory, Integrative and Comparative Physiology. 2011;**300**:R635-R643. DOI: 10.1152/ajpregu.00441.2010

[12] Vlaicu PA, Untea AE, Varzaru I, Saracila M, Oancea AG. Designing nutrition for health—Incorporating dietary by-products into poultry feeds to create functional foods with insights into health benefits, risks, bioactive compounds, food component functionality and safety regulations. Food. 2023;**12**:4001. DOI: 10.3390/foods12214001

[13] Bunchasak C. Role of dietary methionine in poultry production. Journal of Poultry Science. 2009;**46**:169-179. DOI: 10.2141/jpsa.46.169

[14] Selle PH, de Paula Dorigam JC, Lemme A, Chrystal PV, Liu SY. Synthetic and crystalline amino acids: Alternatives to soybean meal in chicken-meat

production. Animals. 2020;**10**:729. DOI: 10.3390/ani10040729

[15] Lingens JB, Abd El-Wahab A, de Paula Dorigam JC, Lemme A, Brehm R, Langeheine M, et al. Evaluation of methionine sources in protein reduced diets for turkeys in the late finishing period regarding performance, footpad health and liver health. Agriculture. 2021;**11**:901. DOI: 10.3390/agriculture11090901

[16] Baker DH, Han Y. Ideal amino acid profile for chicks during the first three weeks posthatching. Poultry Science. 1994;**73**:1441-1447. DOI: 10.3382/ps.0731441

[17] Babazadeh D, Simab PA. Methionine in poultry nutrition: A review. Journal of World's Poultry Science. 2022;**1**:1-11. DOI: 10.58803/jwps.v1i1.1

[18] Liu G, Kim WK. The functional roles of methionine and arginine in intestinal and bone health of poultry: Review. Animals. 2023;**13**:2949. DOI: 10.3390/ani13182949

[19] Jankowski J, Kubińska M, Zduńczyk Z. Nutritional and immunomodulatory function of methionine in poultry diets—A review. Annals of Animal Science. 2014;**14**:17-32

[20] Goulart CC, Costa FGP, Silva JHV, Souza JG, Rodrigues VP, Oliveira CFS. Requirements of digestible methionine + cystine for broiler chickens at 1 to 42 days of age, Exigências de metionina + cistina digestível para frangos de corte de 1 a 42 dias de idade. Revista Brasileira de Zootecnia. 2011;**40**:797-803. DOI: 10.1590/S1516-35982011000400013

[21] Kim D, An B-K, Oh S, Keum M-C, Lee S, Um J-S, et al. Effects of different methionine sources on growth performance, meat yield and blood characteristics in broiler chickens. Journal of Applied Animal Research. 2019;**47**:230-235. DOI: 10.1080/09712119.2019.1617719

[22] Çenesiz AA, Çiftci I, Ceylan N. Effects of DL- and L-methionine supplementation on growth performance, carcass quality and relative bioavailability of methionine in broilers fed maize-soybean-based diets. Journal of Animal and Feed Sciences. 2022;**31**:142-151. DOI: 10.22358/jafs/147800/2022

[23] Lemme A, Hoehler D, Brennan J, Mannion P. Relative effectiveness of methionine hydroxy analog compared to DL-methionine in broiler chickens. Poultry Science. 2002;**81**:838-845. DOI: 10.1093/ps/81.6.838

[24] Knight CD, Atwell CA, Wuelling CW, Ivey FJ, Dibner JJ. The relative effectiveness of 2-hydroxy-4-(methylthio) butanoic acid and DL-methionine in young swine. Journal of Animal Science. 1998;**76**:781-787. DOI: 10.2527/1998.763781x

[25] Rehman AU, Arif M, Husnain MM, Alagawany M, Abd El-Hack ME, Taha AE, et al. Growth performance of broilers as influenced by different levels and sources of methionine plus cysteine. Animals. 2019;**9**:1056. DOI: 10.3390/ani9121056

[26] Macelline SP, Chrystal PV, McQuade LR, Mclnerney BV, Kim Y, Bao Y, et al. Graded methionine dietary inclusions influence growth performance and apparent ileal amino acid digestibility coefficients and disappearance rates in broiler chickens. Animal Nutrition. 2022;**8**:160-168. DOI: 10.1016/j.aninu.2021.06.017

[27] Oshimura E, Sakamoto K. Chapter 19—Amino acids, peptides,

and proteins. In: Sakamoto K, Lochhead RY, Maibach HI, Yamashita Y, editors. Cosmetic Science and Technology. Amsterdam: Elsevier; 2017. pp. 285-303. DOI: 10.1016/B978-0-12-802005-0.00019-7

[28] Nte IJ, Gunn HH, Nte IJ, Gunn HH. Cysteine in broiler poultry nutrition. In: Bioactive Compounds—Biosynthesis, Characterization and Applications. London, UK: IntechOpen; 2021. DOI: 10.5772/intechopen.97281

[29] Elwan HAM, Elnesr SS, Xu Q, Xie C, Dong X, Zou X. Effects of in ovo methionine-cysteine injection on embryonic development, antioxidant status, IGF-I and TLR4 gene expression, and jejunum histomorphometry in newly hatched broiler chicks exposed to heat stress during incubation. Animals (Basel). 2019;**9**:25. DOI: 10.3390/ani9010025

[30] Alagawany M, Elnesr SS, Farag MR, Tiwari R, Yatoo Mohd I, Karthik K, et al. Nutritional significance of amino acids, vitamins and minerals as nutraceuticals in poultry production and health—A comprehensive review. The Veterinary Quarterly. 2021;**41**:1-29. DOI: 10.1080/01652176.2020.1857887

[31] Baker DH. Comparative species utilization and toxicity of sulfur amino acids. The Journal of Nutrition. 2006;**136**:1670S-1675S. DOI: 10.1093/jn/136.6.1670S

[32] Baker DH. Advances in protein-amino acid nutrition of poultry. Amino Acids. 2009;**37**:29-41. DOI: 10.1007/s00726-008-0198-3

[33] Mosharov E, Cranford MR, Banerjee R. The quantitatively important relationship between homocysteine metabolism and glutathione synthesis by the transsulfuration pathway and its regulation by redox changes. Biochemistry. 2000;**39**:13005-13011. DOI: 10.1021/bi001088w

[34] Mack S, Bercovici D, De Groote G, Leclercq B, Lippens M, Pack M, et al. Ideal amino acid profile and dietary lysine specification for broiler chickens of 20 to 40 days of age. British Poultry Science. 1999;**40**:257-265. DOI: 10.1080/00071669987683

[35] Pacheco LG, Sakomura NK, Suzuki RM, Dorigam JCP, Viana GS, Van Milgen J, et al. Methionine to cystine ratio in the total sulfur amino acid requirements and sulfur amino acid metabolism using labelled amino acid approach for broilers. BMC Veterinary Research. 2018;**14**:364. DOI: 10.1186/s12917-018-1677-8

[36] Millecam J, Khan DR, Dedeurwaerder A, Saremi B. Optimal methionine plus cystine requirements in diets supplemented with L-methionine in starter, grower, and finisher broilers. Poultry Science. 2021;**100**:910-917. DOI: 10.1016/j.psj.2020.11.023

[37] Pesti-Asbóth G, Szilágyi E, Molnár PB, Oláh J, Babinszky L, Czeglédi L, et al. Monitoring physiological processes of fast-growing broilers during the whole life cycle: Changes of redox-homeostasis effected to trassulfuration pathway predicting the development of non-alcoholic fatty liver disease. PLoS ONE. 2023;**18**:e0290310. DOI: 10.1371/journal.pone.0290310

[38] Vizzardi E, Bonadei I, Zanini G, Frattini S, Claudia C, Raddino R, et al. Homocysteine and heart failure: An overview. Recent Patents on Cardiovascular Drug Discovery. 2009;**4**:15-21. DOI: 10.2174/157489009787259991

[39] Humberto Vilar Da Silva J, González-Cerón F, Howerth EW,

Rekaya R, Aggrey SE. Inhibition of the transsulfuration pathway affects growth and feather follicle development in meat-type chickens. Animal Biotechnology. 2019;**30**:175-179. DOI: 10.1080/10495398.2018.1461634

[40] Jakubowski H. The role of paraoxonase 1 in the detoxification of homocysteine thiolactone. Advances in Experimental Medicine and Biology. 2010;**660**:113-127. DOI: 10.1007/978-1-60761-350-3_11

[41] Jakubowski H. Homocysteine Modification in Protein Structure/Function and Human Disease. Physiological Reviews. 1 Jan 2019;**99**(1):555-604. DOI: 10.1152/physrev.00003.2018. PMID: 30427275

[42] Włoczkowska O, Perła-Kaján J, Smith D, Jager C, Refsum H, Jakubowski H. Anti-N-homocysteine-protein autoantibodies are associated with impaired cognition. Alzheimer's & Dementia: Translational Research & Clinical Interventions. 2021;**7**:1-10. DOI: 10.1002/trc2.12159

[43] Alirezaei M, Jelodar G, Niknam P, Ghayemi Z, Nazifi S. Betaine prevents ethanol-induced oxidative stress and reduces total homocysteine in the rat cerebellum. Journal of Physiology and Biochemistry. 2011;**67**:605-612. DOI: 10.1007/s13105-011-0107-1

[44] Steenge G, Verhoef P, Katan M. Betaine supplementation lowers plasma homocysteine in healthy men and women. The Journal of Nutrition. 2003;**133**(5):1291-1295. DOI: 10.1093/jn/133.5.1291

[45] Lentz SR. Mechanisms of homocysteine-induced atherothrombosis. Journal of Thrombosis and Haemostasis. 2005;**3**:1646-1654. DOI: 10.1111/j.1538-7836.2005.01364.x

[46] Saande CJ, Pritchard SK, Worrall DM, Snavely SE, Nass CA, Neuman JC, et al. Dietary egg protein prevents hyperhomocysteinemia via upregulation of hepatic betaine-homocysteine S-methyltransferase activity in folate-restricted rats. The Journal of Nutrition. 2019;**149**:1369-1376. DOI: 10.1093/jn/nxz069

[47] Guieu R, Ruf J, Mottola G. Hyperhomocysteinemia and cardiovascular diseases. Annales de Biologie Clinique (Paris). 2022;**80**:7-14. DOI: 10.1684/abc.2021.1694

[48] Baghbanzadeh A, Decuypere E. Ascites syndrome in broilers: Physiological and nutritional perspectives. Avian Pathology. 2008;**37**:117-126. DOI: 10.1080/03079450801902062

[49] Wang Y, Guo Y, Ning D, Peng Y, Cai H, Tan J, et al. Changes of hepatic biochemical parameters and proteomics in broilers with cold-induced ascites. Journal of Animal Science and Biotechnology. 2012;**3**:41. DOI: 10.1186/2049-1891-3-41

[50] Samuels SE. Diet, total plasma homocysteine concentrations and mortality rates in broiler chickens. Canadian Journal of Animal Science. 2003;**83**:601-604. DOI: 10.4141/A02-029

[51] Williams B, Waddington D, Solomon S, Farquharson C. Dietary effects on bone quality and turnover, and Ca and P metabolism in chickens. Research in Veterinary Science. 2000;**69**:81-87. DOI: 10.1053/rvsc.2000.0392

[52] Waqas M, Wang Y, Li A, Qamar H, Yao W, Tong X, et al. Osthole: A coumarin derivative assuage thiram-induced Tibial dyschondroplasia by regulating BMP-2 and RUNX-2

expressions in chickens. Antioxidants. 2019;**8**:330. DOI: 10.3390/antiox8090330

[53] Orth MW, Bai Y, Zeytun IH, Cook ME. Excess levels of cysteine and homocysteine induce tibial dyschondroplasia in broiler chicks. The Journal of Nutrition. 1992;**122**:482-487. DOI: 10.1093/jn/122.3.482

[54] Maharjan P, Hilton K, Weil J, Suesuttajit N, Beitia A, Owens CM, et al. Characterizing woody breast myopathy in a meat broiler line by heat production, microbiota, and plasma metabolites. Frontiers in Veterinary Science. 2020;**6**:1-8

[55] Greene E, Cauble R, Dhamad AE, Kidd MT, Kong B, Howard SM, et al. Muscle metabolome profiles in woody breast-(un)affected broilers: Effects of quantum blue phytase-enriched diet. Frontiers in Veterinary Science. 2020;**7**:458. DOI: 10.3389/fvets.2020.00458

[56] Nutautaitė M, Alijosius S, Bliznikas S, Sasyte V, Viliene V, Pockevičius A, et al. Effect of betaine, a methyl group donor, on broiler chicken growth performance, breast muscle quality characteristics, oxidative status and amino acid content. Italian Journal of Animal Science. 2020;**19**:621-629. DOI: 10.1080/1828051X.2020.1773949

[57] Amin AH, Subbaiah TV, Abbasi KM. Berberine sulfate: Antimicrobial activity, bioassay, and mode of action. Canadian Journal of Microbiology. 1969;**15**:1067-1076. DOI: 10.1139/m69-190

[58] Akhter MH, Sabir M, Bhide NK. Anti-inflammatory effect of berberine in rats injected locally with cholera toxin. The Indian Journal of Medical Research. 1977;**65**:133-141

[59] Bova S, Padrini R, Goldman WF, Berman DM, Cargnelli G. On the mechanism of vasodilating action of berberine: Possible role of inositol lipid signaling system. The Journal of Pharmacology and Experimental Therapeutics. 1992;**261**:318-323

[60] Germoush MO, Mahmoud AM. Berberine mitigates cyclophosphamide-induced hepatotoxicity by modulating antioxidant status and inflammatory cytokines. Journal of Cancer Research and Clinical Oncology. 2014;**140**:1103-1109. DOI: 10.1007/s00432-014-1665-8

[61] Lee YS, Kim WS, Kim KH, Yoon MJ, Cho HJ, Shen Y, et al. Berberine, a natural plant product, activates AMP-activated protein kinase with beneficial metabolic effects in diabetic and insulin-resistant states. Diabetes. 2006;**55**:2256-2264. DOI: 10.2337/db06-0006

[62] Zhao X, Zhang J, Tong N, Chen Y, Luo Y. Protective effects of berberine on doxorubicin-induced hepatotoxicity in mice. Biological & Pharmaceutical Bulletin. 2012;**35**:796-800. DOI: 10.1248/bpb.35.796

[63] Feng X, Sureda A, Jafari S, Memariani Z, Tewari D, Annunziata G, et al. Berberine in cardiovascular and metabolic diseases: From mechanisms to therapeutics. Theranostics. 2019;**9**:1923-1951. DOI: 10.7150/thno.30787

[64] Herrmann M, Taban-Shomal O, Hübner U, Böhm M, Herrmann W. A review of homocysteine and heart failure. European Journal of Heart Failure. 2006;**8**:571-576. DOI: 10.1016/j.ejheart.2005.11.016

[65] McCully KS. Homocysteine and the pathogenesis of atherosclerosis. Expert Review of Clinical Pharmacology. 2015;**8**:211-219. DOI: 10.1586/17512433.2015.1010516

[66] Zhao M, Wang X, He M, Qin X, Tang G, Huo Y, et al. Homocysteine and stroke risk: Modifying effect of methylenetetrahydrofolate reductase C677T polymorphism and folic acid intervention. Stroke. 2017;**48**:1183-1190. DOI: 10.1161/STROKEAHA.116.015324

[67] Li H, He C, Wang J, Li X, Yang Z, Sun X, et al. Berberine activates peroxisome proliferator-activated receptor gamma to increase atherosclerotic plaque stability in *Apoe* $^{-/-}$ mice with hyperhomocysteinemia. Journal of Diabetes Investigation. 2016;**7**:824-832. DOI: 10.1111/jdi.12516

[68] Nygård O, Vollset SE, Refsum H, Stensvold I, Tverdal A, Nordrehaug JE, et al. Total plasma homocysteine and cardiovascular risk profile. The hordaland homocysteine study. JAMA. 1995;**274**:1526-1533. DOI: 10.1001/jama.1995.03530190040032

[69] Werstuck GH, Lentz SR, Dayal S, Hossain GS, Sood SK, Shi YY, et al. Homocysteine-induced endoplasmic reticulum stress causes dysregulation of the cholesterol and triglyceride biosynthetic pathways. The Journal of Clinical Investigation. 2001;**107**:1263-1273. DOI: 10.1172/JCI11596

[70] Woo CWH, Siow YL, Pierce GN, Choy PC, Minuk GY, Mymin D, et al. Hyperhomocysteinemia induces hepatic cholesterol biosynthesis and lipid accumulation via activation of transcription factors. American Journal of Physiology. Endocrinology and Metabolism. 2005;**288**:E1002-E1010. DOI: 10.1152/ajpendo.00518.2004

[71] McCully KS. Vascular pathology of homocysteinemia: Implications for the pathogenesis of arteriosclerosis. The American Journal of Pathology. 1969;**56**:111-128

[72] Chang X, Yan H, Xu Q, Xia M, Bian H, Zhu T, et al. The effects of berberine on hyperhomocysteinemia and hyperlipidemia in rats fed with a long-term high-fat diet. Lipids in Health and Disease. 2012;**11**:86. DOI: 10.1186/1476-511X-11-86

[73] Li Z, Geng Y-N, Jiang J-D, Kong W-J. Antioxidant and anti-inflammatory activities of berberine in the treatment of diabetes mellitus. Evidence-Based Complementary and Alternative Medicine. 2014;**2014**:1-12. DOI: 10.1155/2014/289264

[74] Xie M, Hou SS, Huang W, Fan HP. Effect of excess methionine and methionine hydroxy analogue on growth performance and plasma homocysteine of growing Pekin ducks. Poultry Science. 2007;**86**:1995-1999. DOI: 10.1093/ps/86.9.1995

[75] Lu J, Weil JT, Maharjan P, Manangi MK, Cerrate S, Coon CN. The effect of feeding adequate or deficient vitamin B6 or folic acid to breeders on methionine metabolism in 18-day-old chick embryos. Poultry Science. 2021;**100**:101008. DOI: 10.1016/j.psj.2020.12.075

[76] Ganson FM, Pillai P, Emmert JL. Impact of dietary changes on hepatic homocysteine metabolism in young broilers. Discovery, The Student Journal of Dale Bumpers College of Agricultural, Food and Life Sciences. 2004;**5**:21-26

[77] Kettunen H, Peuranen S, Tiihonen K, Saarinen M. Intestinal uptake of betaine in vitro and the distribution of methyl groups from betaine, choline, and methionine in the body of broiler chicks. Comparative Biochemistry and Physiology Part A: Molecular & Integrative Physiology. 2001;**128**:269-278. DOI: 10.1016/S1095-6433(00)00301-9

[78] Sahebi-Ala F, Hassanabadi A, Golian A. Effect of replacement different methionine levels and sources with betaine on blood metabolites, breast muscle morphology and immune response in heat-stressed broiler chickens. Italian Journal of Animal Science. 2021;**20**:33-45. DOI: 10.1080/1828051X.2020.1868358

[79] Kidd MT, Ferket PR, Garlich JD. Nutritional and osmoregulatory functions of betaine. World's Poultry Science Journal. 1997;**53**:125-139

[80] Simon J. Choline, betaine and methionine interactions in chickens, pigs and fish including crustaceans. World's Poultry Science Journal. 1999;**55**:353-374

[81] Eklund M, Bauer E, Wamatu J, Mosenthin R. Potential nutritional and physiological functions of betaine in livestock. Nutrition Research Reviews. 2005;**18**:31-48. DOI: 10.1079/NRR200493

[82] Abd, El-Ghany WA, Babazadeh D. Betaine: A potential nutritional metabolite in the poultry industry. Animals. 2022;**12**:2624. DOI: 10.3390/ani12192624

[83] Mostashari-Mohases M, Sadeghi AA, Ahmadi J, Esmaeilkhanian S. Effect of betaine supplementation on performance parameters, betaine-homocysteine S-methyltransferase gene expression in broiler chickens consume drinking water with different total dissolved solids. Kafkas Universitesi Veteriner Fakultesi Dergisi. 2017. pp. 563-569. DOI: 10.9775/kvfd.2016.17289

[84] Maidin MBM, McCormack HA, Wilson PW, Caughey SD, Whenham N, Dunn IC. Dietary betaine reduces plasma homocysteine concentrations and improves bone strength in laying hens. British Poultry Science. 2021;**62**:573-578. DOI: 10.1080/00071668.2021.1883550

[85] Sakomura N, Barbosa N, Longo F, Silva ED, Bonato M, Fernandes J. Effect of dietary betaine supplementation on the performance, carcass yield, and intestinal morphometrics of broilers submitted to heat stress. Revista Brasileira de Ciencia Avicola. 2013;**15**:105-112. DOI: 10.1590/S1516-635X2013000200005

[86] Al-Hameed S, Al Machi AS, Al-Gharawi JK. Effect of supplementing betaine on productive performance of broiler chickens fed diets containing different levels of choline. Agricultural and Food Science. 2020. pp. 1649-1653

[87] Wen C, Chen R, Chen Y, Ding L, Wang T, Zhou Y. Betaine improves growth performance, liver health, antioxidant status, breast meat yield, and quality in broilers fed a mold-contaminated corn-based diet. Animal Nutrition. 2021;**7**:661-666. DOI: 10.1016/j.aninu.2020.11.014

[88] Liu Y, Sun B, Zhao A. Effects of dietary betaine on growth performance, digestive function, carcass traits, and meat quality in indigenous yellow-feathered broilers under long-term heat stress. Animals. 2019;**9**:506. DOI: 10.3390/ani9080506

[89] Song Y, Chen R, Yang M, Liu Q, Zhou Y, Zhuang S. Dietary betaine supplementation improves growth performance, digestive function, intestinal integrity, immunity, and antioxidant capacity of yellow-feathered broilers. Italian Journal of Animal Science. 2021;**20**:1575-1586. DOI: 10.1080/1828051X.2021.1986681

[90] Yang Z, Yang JJ, Zhu PJ, Han HM, Wan XL, Yang HM, et al. Effects of betaine on growth performance, intestinal health, and immune response of goslings challenged with lipopolysaccharide. Poultry Science.

2022;**101**:102153. DOI: 10.1016/j.psj.2022.102153

[91] Lint-Kessler CR. Effects of Betaine Supplementation on Growth Performance, Carcass Characteristics, and Blood Parameters of Broilers Reared under Thermoneutral or Heat Stress Conditions. [Master thesis] The University of Tennessee; 2000

[92] Alam B. The Productivity of Broiler Chickens Fed Betaine Supplemented Diets. [Master thesis] Chattogram Veterinary and Animal Sciences University; 2020

[93] Imenshahidi M, Hosseinzadeh H. Berberis vulgaris and berberine: An update review. Phytotherapy Research. 2016;**30**:1745-1764. DOI: 10.1002/ptr.5693

[94] Zhang M, Chen L. Berberine in type 2 diabetes therapy: A new perspective for an old antidiarrheal drug? Acta Pharmaceutica Sinica B. 2012;**2**:379-386. DOI: 10.1016/j.apsb.2012.06.004

[95] Ghavipanje N, Fathi Nasri MH, Vargas-Bello-Pérez E. An insight into the potential of berberine in animal nutrition: Current knowledge and future perspectives. Journal of Animal Physiology and Animal Nutrition. 2023;**107**:808-829. DOI: 10.1111/jpn.13769

[96] Chubarov AS. Homocysteine thiolactone: Biology and chemistry. Encyclopedia. 2021;**1**:445-459. DOI: 10.3390/encyclopedia1020037

[97] Zhang HY, Piao XS, Zhang Q, Li P, Yi JQ, Liu JD, et al. The effects of *Forsythia suspensa* extract and berberine on growth performance, immunity, antioxidant activities, and intestinal microbiota in broilers under high stocking density. Poultry Science. 2013;**92**:1981-1988. DOI: 10.3382/ps.2013-03081

[98] Malekinezhad P, Ellestad LE, Afzali N, Farhangfar SH, Omidi A, Mohammadi A. Evaluation of berberine efficacy in reducing the effects of aflatoxin B1 and ochratoxin A added to male broiler rations. Poultry Science. 2021;**100**:797-809. DOI: 10.1016/j.psj.2020.10.040

[99] Malekinezhad P, Afzali N, Farhangfar SH, Omidi A, Mohammadi A. Berberine improves meat quality and carcass traits in broilers challenged with mycotoxins. Archives of Medical Laboratory Sciences. 2020;**6**:1-9 (e23). DOI: 10.22037/amls.v6.33330

[100] Shen YB, Piao XS, Kim SW, Wang L, Liu P. The effects of berberine on the magnitude of the acute inflammatory response induced by *Escherichia coli* lipopolysaccharide in broiler chickens. Poultry Science. 2010;**89**:13-19. DOI: 10.3382/ps.2009-00243

[101] Yuan L, Li M, Qiao Y, Wang H, Cui L, Wang M. The impact of berberine on intestinal morphology, microbes, and immune function of broilers in response to necrotic enteritis challenge. BioMed Research International. 2021;**2021**:1-9. DOI: 10.1155/2021/1877075

[102] Fernandez CP, Afrin F, Flores RA, Kim WH, Jeong J, Kim S, et al. Downregulation of inflammatory cytokines by berberine attenuates *Riemerella anatipestifer* infection in ducks. Developmental & Comparative Immunology. 2017;**77**:121-127. DOI: 10.1016/j.dci.2017.07.027

Chapter 5

Enzymes in Poultry Feed

Mohamed I. Alshelmani, Salah A. El-Safty, Majdi A. Kairalla and Ali M. Humam

Abstract

Since the use of non-traditional feedstuffs has become more popular in poultry production, the use of exogenous enzymes has become more crucial. In order to lower the cost of ration formulation, low protein diets and unconventional feedstuffs are now being used. Therefore, enzyme supplementation or fermented feedstuffs could release certain nutrients and increase their availability. In conclusion, the supplementation of exogenous enzymes may introduce a positive development in terms of poultry nutrition. For instance, it has been discovered that phytase supplementation may release phosphorus from phytate and reduce phosphorus excretion in broiler manure. In addition, fiber-degrading enzymes have been proven to improve broiler performance and reduce intestinal viscosity. Likewise, protein-degrading enzymes are beneficial in low-protein diets, as they decrease anti-nutritional factors in soybean meal, increase crude protein, amino acids digestibility and reduce nitrogen excretion and ammonia emission in broiler manure, which positively impacts the environment. The supplementation of mixed exogenous enzymes to broiler feed may lead to better utilization of the nutrients on behalf of the chickens. This chapter discusses the most common enzymes in the field of poultry production, such as β-glucanase, xylanase, mannanase, phytase, and protease.

Keywords: enzymes, non-traditional feedstuffs, non-starch polysaccharides, fiber-degrading enzymes, protein-degrading enzymes, poultry feed

1. Introduction

Soybean meal (SBM), which is a major source of protein, and yellow corn, which is a major source of energy, are the two feed components that are most frequently utilized in animal rations. However, the global demand of poultry products (meat and eggs) is rising, particularly in developing countries to cover the gap of protein shortage [1, 2]. In addition, the global population is expected to reach 9.1 billion habitants by the year 2050, and the current trend nowadays is to produce biofuel from feed ingredients, which can create a serious food security threat, especially in the developing regions [1].

As a result, there is considerable interest in incorporating non-traditional feed ingredients in poultry rations to substitute some of the SBM and yellow corn [1] or in using some medicinal plants in poultry diets [3, 4]. Natural alternatives to sub-therapeutic antimicrobials are increasingly being used to improve the performance and safety of broiler products. Many feed additives, like enzymes, as a result, that can reduce the risk of digestive diseases while also improving performance are valuable tools for poultry nutritionists. Nevertheless, the non-traditional feedstuffs have anti-nutritional factors (ANF) or significant amounts of insoluble fiber (cellulose) and non-starch polysaccharides (NSP) as soluble fiber in poultry feed [5]. The low portions of dietary fibers and NSP in poultry diets could be beneficial in terms of gut health [6]. However, high levels of NSP may cause excreta to become more viscous and reduce nutrient availability. Thus, the formulation of poultry feed is constrained by these ANF [5, 7, 8]. Enzymes, by definition, are chemicals or catalysts released by cells to speed up specific chemical reactions. This definition accounts for enzymes released in the digestive tract to aid in the digestion of food. Today, these same enzymes can be effectively manufactured and added to animal feeds. Three classes of enzymes (phytases, carbohydrases, and proteases) are typically considered for use in poultry feeds [9]. Therefore, the supplementation of enzymes in poultry feed could be essential to enhancing digestion and nutrient availability, particularly for young birds. Based on the fact provided above, this chapter discusses the concept of enzyme supplementation to poultry feed and their effects on productive efficiency. Supplementing broiler diets with combinations of xylanase, amylase, and protease has been extensively researched. They have been shown to improve nutrient digestibility and growth performance [10–14]. A combination of amylase, xylanase, and protease enzymes could effectively act together to cleave various bond types in indigestible portions of feed ingredients, leading to increased levels of energy available for growth and/or egg production. The supplementation of these three enzymes to the diet in combination at 500 mg/ton typically increases energy availability to the birds by 3 to 5% [15]. Supplementation of 300 or 600 g/kg diet of such enzymes to broiler or turkey-fed wheat-distillers' dried grains with soluble-based diet showed improvement in metabolizable energy of up to 203 kcal/kg dry matter [10].

2. The significant impact of enzyme supplementation to poultry feed

Because of the increasing consideration in using non-traditional feedstuffs in poultry diets and their limitation in monogastrics, the significance of enzyme supplementation was considerable for researchers. Besides, attention has been paid to the solid-state fermentation by fiber-degrading microbes [16–20].

The increasing price of SBM and yellow corn, which are the main feed ingredients in poultry diets, prompted researchers to consider alternative feed ingredients to face the shortage of the aforementioned feedstuffs. The limitation of using the local feedstuffs or agro-industrial waste in poultry feed is a barrier due to their content of ANF, fibers, and NSP [21]. Hereby, the significance of enzyme supplementation becomes apparent in order to alleviate the negative effect of ANF and to increase the nutrient availability to birds. As a result, the supplementation of enzymes allows the feed manufacturer to be more flexible in using a variety of local raw materials. Moreover, a decrease in the excretion of phosphorus and nitrogen to the field has positive effects on the environment and its elements [22].

The type of enzyme that can be used in poultry feed depends on the substrates or the chemical component of the non-traditional feed material that is used as an alternative to yellow corn. There are many types of enzymes that can be utilized in poultry feed. For instance, if palm kernel cake (PKC) is utilized as chicken feed, β-mannosidase [23], β-mannanase, xylanase, and β-glucanase would be included in the feed [5, 17, 20]. On the other hand, if barley or wheat is utilized as chicken feed, xylanase can break down the arabinoxylans in wheat, and β-glucanase can hydrolyze the β-glucosidic bonds of β-glucans in barley [24]. A research study carried out on broiler fed with diet containing 1% prilled palm fat with lyso-lecethin showed significant enhancement in nutrient digestibility, BWG, and FCR during the experiment [25]. Another important point to consider is that wheat and barley can be fed to young birds up to 40 and 30%, respectively, with the supplementation of enzymes [26].

About 80% of birds' diets are made up of ingredients from plant origin containing non-starch polysaccharides (NSP) in the plant's cell wall. Among NSP, β-mannans can be considered as the leading molecules and are the most prevalent in a wide variety of feed ingredients including soybean meal, which is the major protein source in feeds produced around the world [27]. In practice, poultry diet supplementation with exogenous enzymes is a universal strategy to improve nutrient utilization and growth performance, thus reducing feed cost [28]. In energy-deficient diets (less than 80 kcal/kg from basal diet), the supplementation of β-mannanase enzyme at 250 or 300 g/ton improved growth performance ($P < 0.05$) in broiler from 3 to 5 weeks of age. Accordingly, β-mannanase enzyme supplementation should be considered when low-energy diets are formulated in broiler [29].

3. Fiber-degrading enzymes in broiler feed

Supplementation of enzymes to poultry feed, nowadays, is more considerable by nutritionists to improve the nutritional value to the agro-industrial waste. Thus, it can replace a reasonable portion of yellow corn and SBM in poultry diets.

Numerous studies have been done on broiler chickens to determine the influence of enzyme supplementation on productive efficiency. A study carried out by Kocher et al. [30] showed a significant ($P < 0.05$) improvement in apparent metabolizable energy (AME) and protein digestibility in broilers supplemented with endo-1,3(4)-β –glucanase, hemicellulose, and pectinase at 365 g/kg to their yellow corn and SBM-based diet.

Ng and Chong [31] pointed that fish fed with 40% PKC-based diet and supplemented by exogenous enzyme exhibited improvement in dry matter and energy digestibility. Similar outcomes were reported by Iyayi and Davies [32], who mentioned that supplementation of 0.01% avozyme® in broiler fed with 30% PKC-based diet, led to a significant ($P < 0.05$) increase in feed intake and body weight gain (BWG) during the starter phase. In addition, substantial increase ($P < 0.05$) was observed in apparent digestibility of crude protein, fiber, and fat. However, the carcass characteristics and internal organs were not affected by the supplementation of enzyme.

It was reported that gamanase inclusion (hemicell mannanase from *Bacillus lentus* and mannanase from *Aspergillus niger*) improved the BWG of broiler fed with diet containing PKC [33]. Furthermore, β-mannanase, β-mannosidase, and β-glucanase are the main enzymes that may degrade the mannan molecule [23]. Additional enzymes were suggested by Moreira and Filho [23] to aid the process of degrading mannan such as acetyl mannan esterase and α-galactosidase to cleave the side chain that may be attached to the mannan.

The fiber content in PKC fermented by cellulolytic and hemicellulolytic bacteria showed significant decrease ($P < 0.05$) in neutral detergent fiber (NDF), acid detergent fiber (ADF), crude fiber, cellulose, and hemicellulose [17]. In a digestibility trial conducted by Alshelmani et al. [34] on broiler chickens, the amino acid content and availability of PKC fermented by *Paenibacillus polymyxa* ATCC842 and *P. curdlanolyticus* DSMZ 10248 were significantly ($P < 0.05$) increased (**Tables 1** and **2**).

A feeding trial conducted by Soltan [35] on broiler chickens found that supplementation of enzyme improved the BWG and feed conversion ratio (FCR) in the group fed with 20% PKC-based diet compared to the control group. The

Nutrient (g/kg)	PKC	FPKCa[1]	FPKCb[2]	SEM[3]	P-values
Crude protein	164.3^{b}	168.0^{a}	166.8^{a}	0.04	0.0003
Dry matter	914.2	926.2	924.4	0.38	0.5228
Ash	47.4	46.7	48.0	0.13	0.2201
Crude fiber	169.6^{a}	140.9^{b}	142.9^{b}	0.19	<0.0001
Neutral detergent fiber (NDF)	822.9^{a}	717.0^{b}	735.4^{b}	0.52	<0.0001
Acid detergent fiber (ADF)	514.8^{a}	472.7^{b}	474.5^{b}	0.58	0.0003
Hemicellulose	308.1^{a}	244.3^{b}	264.2^{b}	0.75	0.0010
Cellulose	355.5^{a}	318.5^{b}	314.1^{ab}	0.62	0.0010
In dispensable amino acids					
Lysine	3.7	4.1	3.8	0.02	0.1325
Leucine	8.9	9.4	9.5	0.02	0.0551
Isoleucine	5.0^{b}	5.9^{a}	5.3^{a}	0.02	0.0239
Valine	6.9	7.8	7.2	0.03	0.1433
Phenyl alanine	5.7^{b}	6.6^{a}	6.3^{ab}	0.02	0.0192
Threonine	4.1^{b}	5.1^{a}	4.6^{ab}	0.02	0.0118
Histidine	2.3^{b}	2.9^{a}	2.4^{ab}	0.02	0.0150
Methionine	2.2^{b}	2.7^{a}	2.6^{a}	0.01	0.0003
Arginine	16.0^{b}	17.6^{a}	16.9^{ab}	0.04	0.0312
Glycine	6.0^{b}	7.8^{a}	7.1^{ab}	0.04	0.0489
Dispensable amino acids					
Aspartic acid	11.2^{b}	12.7^{a}	12.3^{ab}	0.03	0.0155
Glutamic acid	24.8^{b}	28.0^{a}	27.6^{a}	0.08	0.0033
Proline	4.4^{b}	5.9^{a}	5.2^{ab}	0.02	0.0018
Serine	5.6^{b}	6.9^{a}	6.6^{ab}	0.04	0.0150
Tyrosine	2.5	2.4	2.4	0.01	0.4435
Cysteine	2.0	2.2	2.1	0.01	0.3632
Alanine	6.2	7.0	7.1	0.06	0.3892

[1]*FPKCa; fermented palm kernel cake by P. polymyxa ATCC 842.*
[2]*FPKCb; fermented palm kernel cake by P. curdlanolyticus DSMZ 10248.*
[3]*Pooled standard error of means.*
[*a,b]*Means ± SEM. Means with different superscripts in the same row are differ significantly ($P < 0.05$).*
n = 6 (6 replicates per treatment with 2 birds per replicate).
A = Adapted from Alshelmani et al. [34].

Table 1.
*Nutrient content of palm kernel cake and fermented palm kernel cake by cellulolytic bacteria (dry matter basis)*A.*

Nutrient (%)	PKC	FPKCa[1]	FPKCb[2]	SEM[3]	P-values
Crude protein	57.92[b]	61.83[a]	60.88[a]	0.63	0.0014
In dispensable amino acids					
Lysine	65.94	69.57	70.63	2.12	0.1479
Leucine	65.47	68.04	63.89	1.39	0.1375
Isoleucine	69.59	70.47	66.39	2.58	0.5152
Valine	62.89[b]	70.42[a]	65.08[b]	1.26	0.0022
Phenyl alanine	68.77	70.76	68.51	1.96	0.6802
Threonine	61.38	64.98	61.69	1.73	0.2935
Histidine	56.99[b]	71.50[a]	64.83[ab]	2.77	0.0076
Methionine	61.67[b]	71.92[a]	69.20[a]	0.90	<0.0001
Arginine	75.75[b]	81.15[a]	76.30[b]	0.95	0.0019
Glycine	47.44	45.52	52.96	4.16	0.4424
Dispensable amino acids					
Aspartic acid	56.87[b]	64.30[a]	61.74[a]	1.20	0.0018
Glutamic acid	62.64[b]	72.37[a]	65.45[b]	0.91	<0.0001
Proline	53.76	58.73	51.20	3.06	0.2401
Serine	65.76	69.78	67.58	2.10	0.4186
Tyrosine	59.04[b]	67.58[a]	61.93[ab]	1.84	0.0155
Cysteine	33.34[b]	41.45[a]	37.46[ab]	2.01	0.0393
Alanine	52.07[b]	66.87[a]	59.84[ab]	2.49	0.0029

[1]*FPKCa; fermented palm kernel cake by P. polymyxa ATCC 842.*
[2]*FPKCb; fermented palm kernel cake by P. curdlanolyticus DSMZ 10248.*
[3]*Pooled standard error of means.*
[*a,b]*Means ± SEM. Means with different superscripts in the same row differ significantly (P < 0.05).*
n = 6 (6 replicates per treatment with 2 birds per replicate).
A = Adapted from Alshelmani et al. [34].

Table 2.
Amino acid and crude protein digestibility of palm kernel cake and fermented palm kernel cake by cellulolytic bacteria (dry matter basis)[A].*

supplementation of 0.015% roxazyme® to broiler feed increased BWG and final body weight. Moreover, the supplementation of such enzyme to the PKC led to a significant (P < 0.05) increase in crude protein from 12 to 17.8%. On the other hand, the crude fiber was significantly (P < 0.05) decreased from 20.2 to 17.3% [36].

The nutritional value of the PKC treated with enzyme [36] or fermented by cellulolytic bacteria [7, 17, 34] improved when compared against untreated PKC. It has been found that fungal growth on lignocellulosic fibers during solid state fermentation decreased NDF, ADF, and hemicellulose. In addition, the ANF (phytate and tannins) declined in some agro-industrial waste as a result of microbial fermentation [37]. Regarding the ANF in animal feed ingredients, it has been found that NSP has been reported to be the main reason affecting nutrient digestibility and increase intestinal viscosity. Therefore, the immunity status and gut microflora will be adversely affected as a result of decreasing nutrient absorption and utilization by the animal [5, 7, 20, 38]. The research demonstrated that supplementation of β-mannanase [38] or solid-state fermentation technique by cellulolytic microorganisms [5, 7, 20, 34] to β-mannan-rich diets may increase nutrient digestibility, enhance immunity status of the bird, increase beneficial microflora in small intestines, and improve the productivity of poultry.

In vitro trial conducted by Zamani et al. [39] showed that *P. polymyxa* ATCC842 and *P. curdlanolyticus* DSMZ 10248 were capable of producing cellulase, xylanase, and mannanase in PKC, rice bran, and wheat pollard. On the other hand, the supplementation of glucanase and xylanase in broiler fed with wheat-based diet improved the gut microflora [40]. The BWG and FCR improved in broiler chickens fed with diet containing 15% barley and supplemented with β-glucanase [41].

4. Keratin-degrading enzymes

The increase in production of poultry around the world resulted in massive waste output. The most poultry waste resulting from poultry processing in slaughterhouses are feathers. Several million tons of such industrial by-products have been recorded [42, 43]. It is reported that feather constitutes about 8% of the adult bird [44] and contains about 85% crude protein [1]. The feather's protein is keratin, and the degradable protein is difficult. Feather meal contains 5% cysteine and 3000 kcal/Kg metabolizable energy. The digestible cysteine is about 60% based on the processing conditions [1].

4.1 Degradation of feather meal

The biological value of a feather meal is low because of its nutrient availability to the animal. However, the fermentation process by microorganisms or using keratinolytic enzyme could improve the nutritional value of such a product [1, 44]. Several keratinases were generated from *Bacillus* spp., *B. licheniformis*, *B. pumilus* [44] *B. subtilis,* and *Aspergillus fumigatus* [1]. It was observed that keratinase supplementation increased amino acid digestibility in raw feather meal from 30 to 66% [44].

The incubation of keratinase from *B. pumilus* A1 at 45 to 60°C for 6 h led to the successful degradation of the feather meal. Therefore, the treated feather meal or even the fermented one can be utilized as an animal feed ingredient [44, 45]. Additionally, fermentation with *B. licheniformis* at 50°C for 5 days may produce a fermented feather meal comparable to that of SBM [1, 46].

Reference	Method	Inclusion of feather meal	Output
Adejumo and Adetunji [47]	Feather meal was fermented by *B. subtilis* to produce microbial biodegraded feather meal.	6%	Improved growth performance
Xu et al. [48]	Supplementation of 200,000 U/kg of keratinase on broiler diet	4%	Improved growth performance, meat quality, and nutrient digestibility
Lee et al. [49]	Feather meal was mixed with soybean meal and fermented by *B. amylolequefaciens* CU33	5%	Improve duodenal morphology and promote digestion and absorption

Table 3.
Effect of keratinase supplementation or fermented feather meal on broiler performance.

As can be seen, the inclusion of feather meal in poultry rations is about 2–3%. Nevertheless, the fermented product by keratin-degrading microbes or keratinase supplementation may provide additional value to such a product, lowering the cost of poultry feed (**Table 3**) [1].

5. Enzyme supplementation on plant protein meals

The most protein-rich source in poultry nutrition is SBM. Raw SBM contains some ANF, such as trypsin inhibitors [46] and lectins [50]. Fortunately, these ANF can be minimized by heating. However, excessive heat leads to decreased lysine availability because lysine is very sensitive to Maillard reaction so that the reducing sugars (raffinose and stachyose) react with the epsilon amino group of lysine and become unavailable [51].

There is a tendency to use low-protein diet in poultry production [46]. The benefits behind that are to reduce nitrogen excretion and ammonia emission from poultry manure to the environment. Additionally, it decreases the cost of feed, increasing the revenue from the production of broilers [52]. Therefore, enzyme supplementation or fermentation processes are being used to break down plant protein for monogastrics. It is recommended to supply protease and phytase to SBM to improve amino acid availability [53] and release more phosphorus from phytate [54].

Phytate molecules can reduce amino acid digestibility by binding dietary amino acids. Therefore, the supplementation of phytase to the chickens increases the availability of phosphorus and amino acids as well [52]. At the same time, it can play an important role in reducing the release of phosphorus into the soil [46]. Other minerals can be increased along with phosphorus as a result of phytase supplementation, so that the availability of other elements, such as zinc, from yellow corn and SBM increases up to 10% [46].

In a feeding trial conducted by Maqsood et al. [55], broiler chickens (Ross 308) fed with low-protein diets (20% of crude protein less than standard allowances) and supplemented with protease at 200 g/ton led to improved growth performance, intestinal health, and carcass characteristics. Similar findings were observed by Tajudeen et al. [56] when birds were fed low-protein diets and administered with 0.022% protease; the birds exhibited an improvement in BWG, crude protein digestibility, and gut morphology. McCafferty et al. [57] reported that protease supplementation in broiler diets improved their growth performance.

Encouraging results were observed among laying hens fed with corn and SBM-based diets and supplemented with protease. A study conducted by Poudel et al. [58] showed that protease supplementation considerably improved crude protein digestibility and increased egg production in laying hens. Wealleans et al. [59] claimed that multi-protease enzyme supplementation to broiler chicken-fed low-protein diets led to enhanced FCR, carcass yield, and gut health. It is suggested that phytase and protease supplementation to low-protein diets can improve crude protein and majority-of-amino acid digestibility [52].

The beneficial impact of protease supplementation on SBM could be attributed to the reduction of ANF, such as trypsin and chymotrypsin inhibitors [56]. The mechanism can occurred via the release of more peptides from ANF that exist in SBM (**Table 4**) [58].

Supplemented enzyme	Influence	Reference
Protease at 200 g/ton to low-protein diet (20% reduction of protein from standard requirements of Ross 308).	Improved growth performance, gut health, and carcass traits	Maqsood et al. [55]
Protease at 0.022% to low-protein diet (0.75% lower than standard requirements).	Optimum findings in BWG and nutrient digestibility	Tajudeen et al. [56]
Phytase, xylanase, and protease at 2000 U/kg, 200 U/kg, and 15,000 U/kg, respectively.	Improved broiler performance	McCafferty et al. [57]
Protease at 60 g/ton of feed.	Increased egg income and return on investment.	Poudel et al. [58]
Multi protease at 300 mg/kg diet to low protein diet (3.5% lower than standard requirements of Ross 308).	Improved FCR, carcass weight and yield, breast yield, and gut health and morphology.	Wealleans et al. [59]

Table 4.
Effect of diet supplemented with protease and phytase on broiler performance.

6. Conclusion

- In conclusion, the supplementation of exogenous enzymes to poultry feed may introduce a positive development in terms of poultry nutrition. For instance, it has been discovered that phytase supplementation to broiler diets may release phosphorus from phytate and reduce phosphorus excretion in broiler manure.

- In addition, fiber-degrading enzymes have also been proven to improve broiler performance and reduce intestinal viscosity. Likewise, protein-degrading enzymes are also beneficial in low-protein diets, so that it decreases ANF in SBM, releases amino acids, increases crude protein and most-amino-acid digestibility, and reduces nitrogen excretion and ammonia emission in broiler manure, which positively impacts the environment.

- The supplementation of mixed exogenous enzymes to broiler feed may lead to better utilization of the nutrients on behalf of the chickens.

- The dosage of enzyme in poultry feed depends on the enzyme activity and the manufacturer recommendation.

Author details

Mohamed I. Alshelmani[1]*, Salah A. El-Safty[2], Majdi A. Kairalla[3] and Ali M. Humam[4]

1 Faculty of Agriculture, Deptartment of Animal Production, University of Benghazi, Benghazi, Libya

2 Faculty of Agriculture, Department of Poultry Production, Ain Shams University, Hadayek Shoubra, Cairo, Egypt

3 Faculty of Agriculture, Deptartment of Animal Production, University of Sebha, Sebha, Libya

4 Animal Production Department, Agricultural Engineering Sciences College, University of Baghdad, Baghdad, Iraq

*Address all correspondence to: mohammed.alshelmani@uob.edu.ly

References

[1] Alshelmani MI, Abdalla EA, Kaka U, Basit MA. Nontraditional feedstuffs as an alternative in poultry feed. In: Advances in Poultry Nutrition Research. London: IntechOpen; 2021

[2] Kareem KY, Abdulla NR, Foo HL, Mohd AN, Zamri NS, Loh TC, et al. Effect of feeding larvae meal in the diets on growth performance, nutrient digestibility and meat quality in broiler chicken. Indian Journal of Animal Sciences. 2018;**88**(10):1180-1185

[3] Kairalla MA, Aburas AA, Alshelmani MI. Effect of diet supplemented with graded levels of ginger (*Zingiber officinale*) powder on growth performance, hematological parameters, and serum lipids of broiler chickens. Archives of Razi Institute. 2022;**77**(6):2077-2083

[4] Kairalla MA, Alshelmani MI, Aburas AA. Effect of diet supplemented with graded levels of garlic (*Allium sativum* L.) powder on growth performance, carcass characteristics, blood hematology, and biochemistry of broilers. Open Veterinary Journal. 2022;**12**(5):595-601

[5] Alshelmani MI, Kaka U, Abdalla EA, Humam AM, Zamani HU. Effect of feeding fermented and non-fermented palm kernel cake on the performance of broiler chickens: A review. World's Poultry Science Journal. 2021;**77**(2):377-388

[6] Jha R, Mishra P. Dietary fiber in poultry nutrition and their effects on nutrient utilization, performance, gut health, and on the environment: A review. Journal of Animal Science and Biotechnology. 2021;**12**:1-16

[7] Alshelmani MI, Loh TC, Foo HL, Sazili AQ, Lau WH. Effect of feeding different levels of palm kernel cake fermented by *Paenibacillus polymyxa* ATCC 842 on nutrient digestibility, intestinal morphology, and gut microflora in broiler chickens. Animal Feed Science and Technology. 2016;**216**(6):216-224

[8] Raza A, Bashir S, Tabassum R. An update on carbohydrases: Growth performance and intestinal health of poultry. Heliyon. 2019;**5**(4):e01437

[9] Park CS, Adeola O. Enzymes and enzyme supplementation of swine diets. In: Chiba LI, editor. Sustainable Swine Nutrition. 2022. pp. 445-469

[10] Adebiyi A, Olukosi O. Metabolizable energy content of wheat distillers' dried grains with solubles supplemented with or without a mixture of carbohydrases and protease for broilers and turkeys. Poultry Science. 2015;**94**(6):1270-1276

[11] Liu S, Cadogan D, Péron A, Truong H, Selle P. A combination of xylanase, amylase and protease influences growth performance, nutrient utilisation, starch and protein digestive dynamics in broiler chickens offered maize-, sorghum-and wheat-based diets. Animal Production Science. 2014;**55**(10):1255-1263

[12] Olukosi O, Beeson L, Englyst K, Romero L. Effects of exogenous proteases without or with carbohydrases on nutrient digestibility and disappearance of non-starch polysaccharides in broiler chickens. Poultry Science. 2015;**94**(11):2662-2669

[13] Romero L, Sands J, Indrakumar S, Plumstead P, Dalsgaard S, Ravindran V.

Contribution of protein, starch, and fat to the apparent ileal digestible energy of corn-and wheat-based broiler diets in response to exogenous xylanase and amylase without or with protease. Poultry Science. 2014;**93**(10):2501-2513

[14] Amerah A, Romero L, Awati A, Ravindran V. Effect of exogenous xylanase, amylase, and protease as single or combined activities on nutrient digestibility and growth performance of broilers fed corn/soy diets. Poultry Science. 2017;**96**(4):807-816

[15] Cowieson AJ, Ravindran V. Sensitivity of broiler starters to three doses of an enzyme cocktail in maize-based diets. British Poultry Science. 2008;**49**(3):340-346

[16] Alshelmani MI, Loh TC, Foo HL, Lau WH, Sazili AQ. Characterization of cellulolytic bacterial cultures grown in different substrates. The Scientific World Journal. 2013;**2013**:6

[17] Alshelmani MI, Loh TC, Foo HL, Lau WH, Sazili AQ. Biodegradation of palm kernel cake by cellulolytic and Hemicellulolytic bacterial cultures through solid state fermentation. The Scientific World Journal. 2014;**2014**:8

[18] Alshelmani MI, Loh TC, Foo HL, Sazali AQ, Lau WH. Effect of feeding fermented palm kernel cake on performance of broiler chickens. In: Proceeding WPSA and WVPA (Malaysia Branch) Scientific Conference. Serdang, Malaysia: Universiti Putra Malaysia; 2013. p. 83

[19] Alshelmani MI, Loh TC, Foo HL, Sazili AQ, Lau WH. Effect of feeding different levels of palm kernel cake fermented by *Paenibacillus polymyxa* ATCC 842 on broiler growth performance, blood biochemistry, carcass characteristics, and meat quality. Animal Production Science. 2017;**57**(5):839-848

[20] Alshelmani MMI. Effect of feeding palm kernel cake fermented by fiber degrading bacteria on performance of broiler chicken [Doctor of Philosophy thesis]. Malaysia: UPM; 2015

[21] Zamani HU, Loh TC, Foo HL, Samsudin AA, Alshelmani MI. Effects of feeding palm kernel cake with crude enzyme supplementation on growth performance and meat quality of broiler chicken. International Journal of Microbiology and Biotechnology. 2017;**2**(1):22-28

[22] Khattak F, Pasha T, Hayat Z, Mahmud A. Enzymes in poultry nutrition. Journal of Animal and Plant Science. 2006;**16**(1-2):1-7

[23] Moreira L, Filho E. An overview of mannan structure and mannan-degrading enzyme systems. Applied Microbiology and Biotechnology. 2008;**79**(2):165-178

[24] Ravn JL, Martens HJ, Pettersson D, Pedersen NR. A commercial GH 11 xylanase mediates xylan solubilisation and degradation in wheat, rye and barley as demonstrated by microscopy techniques and wet chemistry methods. Animal Feed Science and Technology. 2016;**219**:216-225

[25] Jaapar M, Alshelmani M, Humam A, Loh T, Foo H, Akit H. Effect of feeding prilled palm fat with lyso-lecithin on broiler growth performance, nutrient digestibility, lipid profile, carcass, and meat quality. Poultry Science Journal. 2020;**8**(1):43-50

[26] Leeson S, Summers J. Commercial Poultry Nutrition. 2nd ed. Guelph, Ontario, Canada: University Books; 1997

[27] Caldas JV, Vignale K, Boonsinchai N, Wang J, Putsakum M, England JA, et al. The effect of β-mannanase on nutrient utilization and blood parameters in chicks fed diets containing soybean meal and guar gum. Poultry Science. 2018;**97**(8):2807-2817

[28] Bedford MR. Exogenous enzymes in monogastric nutrition—Their current value and future benefits. Animal Feed Science and Technology. 2000;**86**(1-2):1-13

[29] Hashim MM, El-Safty SA, El-Eraqi KG, El-Sherif HMR, Azza T, Marwa K, et al. Effect of β-Mannanase enzymes supplementation to energy deficient diets on productive performance, physiological and carcass traits of broilers. International Journal of Poultry Science. 2020;**19**(10):455-466

[30] Kocher A, Choct M, Porter MD, Broz J. Effects of feed enzymes on nutritive value of soyabean meal fed to broilers. British Poultry Science. 2002;**43**(1):54-63

[31] Ng WK, Chong KK. The nutritive value of palm kernel meal and the effect of enzyme supplementation in practical diets for red hybrid tilapia (*Oreochromis* sp.). Asian Fisheries Science. 2002;**15**:167-176

[32] Iyayi EA, Davies BI. Effect of enzyme supplementation of palm kernel meal and brewer's dried grain on the performance of broilers. International Journal of Poultry Science. 2005;**4**(2):76-80

[33] Sundu B, Dingle J. Use of enzymes to improve the nutritional value of palm kernel meal and copra meal. Queensland Poultry Science Symposium Australia. 2002;**11**(14):1-15

[34] Alshelmani MI, Loh TC, Foo HL, Sazili AQ, Lau WH. Effect of solid state fermentation on nutrient content and ileal amino acids digestibility of palm kernel cake in broiler chickens. Indian Journal of Animal Sciences. 2017;**87**(9):1135-1140

[35] Soltan MA. Growth performance, immune response and carcass traits of broiler chicks fed on graded levels of palm kernel cake without or with enzyme supplementation. Livestock Research for Rural Development. 2009;**21**(37):37-55

[36] Lawal TE, Iyayi EA, Adeniyi BA, Adaramoye OA. Biodegradation of palm kernel cake with multienzyme complexes from fungi and its feeding value for broiler. International Journal of Poultry Science. 2010;**9**(7):695-701

[37] Graminha EBN, Gonçalves AZL, Pirota RDPB, Balsalobre MAA, Da Silva R, Gomes E. Enzyme production by solid-state fermentation: Application to animal nutrition. Animal Feed Science and Technology. 2008;**144**(1-2):1-22

[38] Saeed M, Ayaşan T, Alagawany M, El-Hack M, Abdel-Latif M, Patra A. The role of ß-mannanase (Hemicell) in improving poultry productivity, health and environment. Brazilian Journal of Poultry Science. 2019;**21**:1-8

[39] Zamani HU, Loh TC, Foo HL, Samsudin AA, Alshelmani MI. Comparative evaluation of cellulolytic enzyme production by cellulolytic bacteria via solid state fermentation on palm kernel cake, rice bran, and wheat pollard. Scientific Times Journal of Agricultural Science. 2016;**1**(1):1002

[40] Kouzounis D, Kers JG, Soares N, Smidt H, Kabel MA, Schols HA. Cereal type and combined xylanase/glucanase supplementation influence the cecal microbiota composition in broilers. Journal of Animal Science and Biotechnology. 2022;**13**(1):1-12

[41] Toghyani M, Macelline S, Greenhalgh S, Chrystal P, Selle P, Liu S. Optimum inclusion rate of barley in diets of meat chickens: An incremental and practical program. Animal Production Science. 2022;**62**(7):645-660

[42] Sivakumar N, Raveendran S. Keratin degradation by bacteria and fungi isolated from a poultry farm and plumage. British Poultry Science. 2015;**56**(2):210-217

[43] Lakshmi VV, Aruna Devi D, Jhansi Rani KP. Wealth from poultry waste. In: Ghosh SK, Bhattacharya C, Satyanarayana SV, Varadarajan S, editors. Emerging Technologies for Waste Valorization and Environmental Protection. Singapore: Springer Singapore; 2020. pp. 135-144

[44] Brandelli A, Sala L, Kalil SJ. Microbial enzymes for bioconversion of poultry waste into added-value products. Food Research International. 2015;**73**:3-12

[45] Fakhfakh N, Gargouri M, Dahmen I, Sellami-Kamoun A, El Feki A, Nasri M. Improvement of antioxidant potential in rats consuming feathers protein hydrolysate obtained by fermentation of the keratinolytic bacterium, *Bacillus pumilus* A1. African Journal of Biotechnology. 2012;**11**(4):938-949

[46] Leeson S, Summers J. Commercial Poultry Nutrition. 3rd ed. Department of Animal and Poultry Science, University of Guelph; University Book; 2005

[47] Adejumo IO, Adetunji CO. Production and evaluation of biodegraded feather meal using immobilised and crude enzyme from Bacillus subtilis on broiler chickens. Brazilian Journal of Biological Sciences. 2018;**5**(10):405-416

[48] Xu K-L, Gong G-X, Liu M, Yang L, Xu Z-J, Gao S, et al. Keratinase improves the growth performance, meat quality and redox status of broiler chickens fed a diet containing feather meal. Poultry Science. 2022;**101**(6):101913

[49] Lee T-Y, Lee Y-S, Yeh R-H, Chen K-H, Chen K-L. Bacillus amyloliquefaciens CU33 fermented feather meal-soybean meal product improves the intestinal morphology to promote the growth performance of broilers. Poultry Science. 2022;**101**(9):102027

[50] Han X, Sun Y, Huangfu B, He X, Huang K. Ultra-high-pressure passivation of soybean agglutinin and safety evaluations. Food Chemistry: X. 2023;**18**:100726

[51] Oliveira MSF, Wiltafsky MK, Lee SA, Kwon WB, Stein HH. Concentrations of digestible and metabolizable energy and amino acid digestibility by growing pigs may be reduced by autoclaving soybean meal. Animal Feed Science and Technology. 2020;**269**:114621

[52] Woyengo T, Knudsen KB, Børsting C. Low-protein diets for broilers: Current knowledge and potential strategies to improve performance and health, and to reduce environmental impact. Animal Feed Science and Technology. 2023;**297**:115574 (1-18)

[53] Walk C, Pirgozliev V, Juntunen K, Paloheimo M, Ledoux D. Evaluation of novel protease enzymes on growth performance and apparent ileal digestibility of amino acids in poultry: Enzyme screening. Poultry Science. 2018;**97**(6):2123-2138

[54] Erdaw M, Bhuiyan M, Iji P. Enhancing the nutritional value of soybeans for poultry through supplementation with new-generation feed enzymes. World's Poultry Science Journal. 2016;**72**(2):307-322

[55] Maqsood MA, Khan EU, Qaisrani SN, Rashid MA, Shaheen MS, Nazir A, et al. Interactive effect of amino acids balanced at ideal lysine ratio and exogenous protease supplemented to low CP diet on growth performance, carcass traits, gut morphology, and serum metabolites in broiler chicken. Tropical Animal Health and Production. 2022;**54**(3):186

[56] Tajudeen H, Hosseindoust A, Ha S, Moturi J, Mun J, Lee C, et al. Effects of dietary level of crude protein and supplementation of protease on performance and gut morphology of broiler chickens. European Poultry Science/Archiv für Geflügelkunde. 2022;**86**:1-12

[57] McCafferty KW, Morgan NK, Cowieson AJ, Choct M, Moss AF. Varying apparent metabolizable energy concentrations and protease supplementation affected broiler performance and jejunal and ileal nutrient digestibility from 1 to 35 d of age. Poultry Science. 2022;**101**(7):101911

[58] Poudel I, Hodge VR, Wamsley KGS, Roberson KD, Adhikari PA. Effects of protease enzyme supplementation and varying levels of amino acid inclusion on productive performance, egg quality, and amino acid digestibility in laying hens from 30 to 50 weeks of age. Poultry Science. 2023;**102**(3):102465

[59] Wealleans A, Ashour R, Abu Ishmais M, Al-Amaireh S, Gonzalez-Sanchez D. Comparative effects of proteases on performance, carcass traits and gut structure of broilers fed diets reduced in protein and amino acids. Journal of Animal Science and Technology. 2023. DOI: 10.5187/jast.2023.e20

Chapter 6

Acidifiers as Alternatives for Antibiotics Reduction and Gut Health Improvement for Poultry and Swine

Nguyen Vu Thuy Hong Loan, Ho Trung Thong, Le Nu Anh Thu and Ho Viet Duc

Abstract

Using antibiotics of low doses as feed additives could support to improve poultry and swine performances. However, these applications have caused resistance of bacteria and antibiotic residues in foods of animal origins. Therefore, efforts were focused on solutions to replace antibiotics as growth promoters (AGPs). There are many alternatives for AGPs, in which organic acids are one of the important alternatives. The aim of this chapter is to review publications on these acids and their other forms namely as acidifiers using as feed additives including their names and forms, mode of actions, spectrum against bacteria, combinations among them, and latest updates on their effects on swine and poultry production. The scientific findings show that acidifiers can inhibit pathogenic bacteria growth, improve nutrient digestibility, enhance immunity and overall gut health, consequently increase performances of poultry and swine. Several acids and their salts in both liquid and solid forms have been studied and applied as poultry and swine feed additives; however, the efficacy levels and the mode of actions are dependent on the single acidifiers, their salts, and combinations among them. The uses of acidifiers in their salts and derivative forms and mixtures of different acidifiers seem to be more favorable.

Keywords: acidifiers, antibiotics, organic acids, poultry production, swine production

1. Introduction

Antibiotics, since their discovery in the 1920s, have been widely used as antimicrobial growth promoters in animal production to enhance productivity and prevent diseases [1, 2]. However, due to the emerging resistance against microbes and their residues in meat, milk, and egg, the World Health Organization (WHO) published guidance and recommendations to reduce the use of antibiotics in 1997. About a decade later, the European Union imposed a complete ban on the use of prophylactic antibiotics in the animal feedstuff [3, 4]. A withdrawal of growth-promoting antibiotics in livestock production has led to problems like an increase in the incidence of

IntechOpen

animal diseases and a reduction in productivity [5]. Consequently, various alternatives were sought and explored to replace the use of antibiotics in animal production to maintain performance and their health. The potential substitutes to antibiotics include probiotics and prebiotics, plant extracts, essential oils, antimicrobial peptides, functional amino acids, hyperimmune antibodies, clays, metals, and/or organic acids [6–16]. Among these alternatives, dietary organic acids, also known as acidifiers, have been applied worldwide for decades due to their strong antibacterial, anti-fungal, and anti-mold properties [17]. The organic acids with antibacterial activity are either simple monocarboxylic acid such as butyric acid, propionic acid, acetic acid, and formic acid, or carboxylic acid bearing a hydroxyl group such as tartaric acid, citric acid, malic acid, and lactic acid [18]. These are usually weak organic acids that are capable of lowering the pH of the stomach and in the gastrointestinal tract (GIT), thus inhibiting the growth of pathogenic bacteria, promoting proteolytic enzyme activity and nutrient digestibility, creating stability of the microbial population, and stimulating the growth of beneficial bacteria [19]. Single organic acids have been reported to own a wide range of microbial activities such as physiology, pH range, and membrane structure. Thus, the inclusion of organic acids mixtures in diets is not always consistent, and the response to dietary organic acids could be affected by the type of organic acids, dosage, feed formula, and the age of animals [20]. Therefore, the purpose of this review is to summarize recent studies about responses of swine and poultry to both single and a blend of organic acids aiming to support the overall insight about the effective utilization of organic acids in swine and poultry production for enhancing the performance and gut health. In addition, modes of action of organic acids (OAs) and their classification are also discussed.

2. Classification of acidifiers

Acidifiers, or so-called organic acids, are organic compounds that possess acidic properties. In general, acidifiers are divided into three functional groups including short-chain fatty acids (SCFAs, C1 to C5), medium-chain fatty acids (MCFA; C6 to C12), and tricarboxylic acids (TCA) [21]. In which, SCFAs are most commonly used, such as formic acid (C1), acetic acid (C2), propionic acid (C3), lactic acid (C3), and butyric acids (C4) [22]. These SCFAs are produced in the lower intestine of animals by the microbial fermentation of indigestible sugars and amino acids. Their pKa values are small with a range from higher than 3 to less than 5 (**Table 1**). Since this property, they can selectively inhibit the intestinal bacteria, and thus improve intestinal morphology and decrease the intestinal inflammation [23]. MCFAs are also used in combination with SCFAs as feed additive to enhance the activity of acidifiers in GIT. MCFA can disrupt the phospholipid membrane, thus exhibit potent antibacterial activity. The MCFA commonly used in livestock production include caproic acid (C6), caprylic acid (C8), capric acid (C10), and lauric acid (C12). There has been an increase in recent interest in research relevant to inhibitory activity of MCFA against a wide range of pathogens in the swine industry. For example, lauric acid and a mixture of caprylic and capric acids were reported to exhibit antibacterial activity against pathogenic bacteria such as *Escherichia coli*, *Streptococcus suis*, *Salmonella poona,* and *Clostridium perfringens* [24]. TCA is an organic carboxylic acid whose chemical structure contains three carboxyl functional groups (-COOH). They are metabolic intermediates of Krebs cycle or citric acid cycle, thus are involved in the major energy-yielding metabolic

Classification	Name	Used salts and derivates
Short-chain fatty acid (SCFA)	Formic acid	Ammonium formate Sodium di-formate
	Acetic acid	Sodium acetate
	Propionic acid	Ammonium propionate; Sodium propionate
	Lactic acid	Sodium lactate
	Butyric acid	Sodium butyrate mono, di-, tri-butyrin
	Valeric acid	Glyceride esters
	Benzoic acid	Benzoate
	Malic acid	Sodium, calcium-malate
Medium-chain fatty acid (MDFA)	Caproic acid	Caproates, hexanoates, caproate esters
	Lauric acid	Calcium laurate
	Caprylic acid	—
	Capric acid	—
	Sorbic acid	Calcium sorbate Potassium sorbate Sorbic chloride
Tricarboxylic acid (TCA)	Citric acid	Sodium citrate

Table 1.
Common acidifiers used as additives in swine and poultry production.

pathway in cells. These acids improve gut morphology and barrier function with positive influences on intestinal bacteria community. The best-known TCA is citric acid which has been reported that it can be a potential alternative to antibiotics in animal production [25–27].

Moreover, due to difficulties of using organic acids in practice including offensive odor and their inability to affect the lower part of GIT, different forms of organic acids such as their salts and derivatives have been developed and investigated for their effects on growth performances and gut health [28]. For examples, sodium butyrate and butyrate glycerides (mono-, di-, and tri-butyrin) were reported to have positive influences on animal production including enhancement of gut health, control of pathogens, reduction of inflammation, and improvement of performances [29]. The inclusion of valeric acid glyceride ester in the broiler dietary can improve the feed conversion ratio, positively impact to the intestinal morphology, increase the density of glucagon-like peptide-2 immunoreactive cells, and significantly reduce the number of birds infected necrotic enteritis [30]. Besides, owing to the advantages of today's modern technologies, especially encapsulation technology, which has been widely employed across various scientific fields, including animal nutrition, it effectively overcomes the limitations of conventional feeding methods [31, 32]. Coated organic acids with encapsulated nano/micro materials led to an increase in the stability, bioavailability, and their activity. For example, Feye et al. (2020) and Muniyappan et al. (2021) recently reported that the dietary inclusion of microencapsulated blend of organic acids enhanced the GIT microbiota and may be a viable antibiotic alternative for the swine and poultry industry [33, 34].

3. Mode of action

The use of acidifiers and their salts in the diet of swine and poultry with a reasonable dose can increase the body weight (ADG), improve feed conversion ratio (FCR), and reduce the pathogenic bacteria [35, 36]. Thus, it is necessary to explore the activity of acidifiers. Generally, the mechanisms of action of organic acids include: (i) Lowering of intestinal pH; (ii) Improving nutrient digestibility via the reduction of pH value by release of hydrogen ions in the stomach, thereby activating pepsinogen to form pepsin; (iii) Inhibition of Gram-negative bacteria in the gastrointestinal tract (GIT); (iv) Improved energetic utilization in the intermediate metabolism to enhance endogenous enzyme secretion and chelate minerals; (v) intestinal anti-inflammation and immunity response.

3.1 Lowering of intestinal pH

Organic acids are weak acids in the sense that a certain proportion of the molecules do not fully dissociate. These undissociated, uncharged molecules diffuse easily across the bacterial cell membrane to reach the interior of the cell. After the entry of organic acids into the microbial cell, these acids release the proton (H+) in the more alkaline environment of the cytoplasm, causing a drop of bacterial intracellular pH. This impacts on bacterial metabolism, inhibiting the action of important microbial enzymes. The bacterial cell is forced to use energy to expel the protons, leading to an intracellular accumulation of acid anions. The anions within the bacterial cell are thought to disrupt the metabolic processes in the cell, consequently affecting cell multiplication and limiting growth [4, 17, 18, 36]. There are two major types of organic acids that have different modes of action in decreasing pH. The first group including lactic, fumaric, and citric acid lowers the pH of the stomach leading to indirect reduction of the population of acid sensitive bacteria. The second group including butyric, formic, acetic, propionic, and sorbic has ability to lower the pH of the GIT by penetrating the Gram-negative bacteria cell wall and directly controlling the pathogens [28].

3.2 Improving nutrient digestibility and gut morphology

Since organic acids can reduce the pH value in the GIT, thus, pepsinogen is activated to form pepsin, which causes proteolysis of protein. The protein contents are then broken down into simple peptides and amino acids that can be easily absorbed in the small intestine. In addition, in the presence of an acidic environment, bacterial metabolites such as ammonia and amines are reduced, thereby enhancing digestibility. Therefore, organic acid used as an acidifier in swine and poultry production has been considered to be a potential alternative to antibiotics for improving nutrient digestibility. Previous trials have reported that including 0,5% fumaric acid, 0,5% formic acid, 0,75% acetic acid, or 2% citric acid in broiler diets improved ME, crude protein, ether extract, crude fiber, and nitrogen-free extract [37–39]. Similarly, in swine production, the supplementation of 0,1 or 0,2% of coated organic acid including 17% fumaric acid, 13% citric acid, 10% malic acid, and 1.2% MCFA (capric and caprylic acid) in basal diets linearly increased the dry matter, nitrogen, and energy digestibility [40]. Moreover, low pH also increases the digestibility of nutrients via the changes in the villus height and depth in the small intestines, thus improving the gut morphology and is one of the reasons for the improvement of the feed to gain ratio. For example, in a study by Garciá et al. (2007), broilers fed diets containing 0.5 and

1.0% formic acid exhibited longer villi (1273 and 1250 μm, respectively) compared to the control group (1088 μm) [39]. Panda et al. (2009) reported that the addition of 0.2, 0.4, or 0.6% butyrate in the broiler's diet improved the villus length and crypt depth in the duodenum [41], in which, 0,4% of butyric acid supplementation improved performances. Similarly, Galfi and Bokori (1990) showed an increase in the length of microvilli in the ileum and the depths of the crypts in caecum in growing pigs when fed with 0.17% of sodium butyrate. This dietary increased the average daily body mass gain of pigs by 23.5% [42].

3.3 Inhibition of pathogenic bacteria

It is reported that most common bacteria that affect the intestinal health of both poultry and swine are Gram-negative bacteria such as *Escherichia coli*, *Salmonella*, and *Campylobacter* which can be controlled by supplementation of organic acids in diets [43–45]. The study in mode of action of organic acids showed that most of pathogenic bacteria reside at a pH close to 7, while useful bacteria survive better at a pH between 5.8 and 6.2. Therefore, owing to the intestinal pH lowering capable of organic acids, the population of the pathogenic microbes is reduced that do not affect to beneficial bacteria. In addition, the efficacy of an acid in inhibition of the pathogenic bacterial growth is dependent on its pKa value—the pH at which the acid is half dissociated. Organic acids, most of them, with antimicrobial activity, have a pKa between 3 and 5 (**Table 2**).

Organic acids with higher pKa values are commonly used as preservatives for animal feed. Their antimicrobial efficacy depends on the increasing number of carbon chains and unsaturation properties [48]. Peh et al. (2020) recently reported in-vitro susceptibility of *Campylobacter spp* to 10 organic acids including caprylic acid, sorbic acid, caproic acid, benzoic acid, ascorbic acid, propionic acid, acetic acid, formic acid, fumaric acid, and tartaric acid. In which, the antimicrobial activity of caprylic acid and sorbic acid against *Campylobacter spp* at the lowest minimum inhibitory concentration values measured at pH 7.3 ranged from 0 to 2 nmol/L and 1 to 4 nmol/L, respectively [47].

Organic acids	**pKa value**	**Minimum inhibitory concentration (nmol/l)**	
		E. coli	*Campylobacter jejuni*
Acetic	4.75	1.55	64.00
Benzoic	4.19	0.316	8.0
Butyric	4.81	1.41	nd
Citric	3.13	38.2	nd
Formic	3.75	64.0	128.0
Lactic	3.86	3.72	nd
Malic	3.40	50	nd
Propionic	4.87	64.0	32.0
Sorbic	4.76	4.0	4.0

nd: not detected.

Table 2.
The pKa values of common organic acids and the minimum inhibitory concentration (MIC) of these organic acids against pathogenic bacteria [46, 47].

3.4 Provision of energy source in the GIT

Organic acids act as an energy source in the GIT as they are metabolic intermediates from Krebs cycle, thus directly influencing intestinal metabolic status. For example, Kirchgessner and Roth found that fumaric acid, a product of metabolic pathway in the Kreb cycle, can be used as an energy source with an efficiency close to that of glucose in pigs [49]. In addition, the beneficial effects of organic acids on the growth performance were considered due to their energy contribution. Blank et al. reported that fumaric acid as an available energy source can influence the intestinal mucosa and thus increasing the absorptive surface and capacity of the small intestines due to the rapid recovery of the gut epithelial cells of pigs after weaning [50]. Besides, the intestinal microbiota can ferment fibers and oligosaccharides to produce SCFAs including acetate, propionate, and butyrate. These metabolites play a significant role in maintaining the intestinal homeostasis [51]. SCFAs were reported to contribute 5–15% and 60–70% of the total energy requirements of colonic epithelial cells in humans, respectively. Among SCFAs, butyrate is the major energy source for colonocytes, which have beneficial effects on both cellular energy metabolism and intestinal homeostasis [52]. Donohoe et al. also showed that butyrate maintains energy homeostasis and prevents autophagy by acting as an energy source rather than a histone deacetylase inhibitor in mammalian colon [53].

3.5 Preventing the intestinal inflammation status and supporting immunity homeostasis

There is mechanistic evidence for the effects of SCFA on mucosal immune and inflammatory status, based on studies involving cell lines and small animal models [51]. SCFAs, particularly butyrate, have been shown to exert their effects through

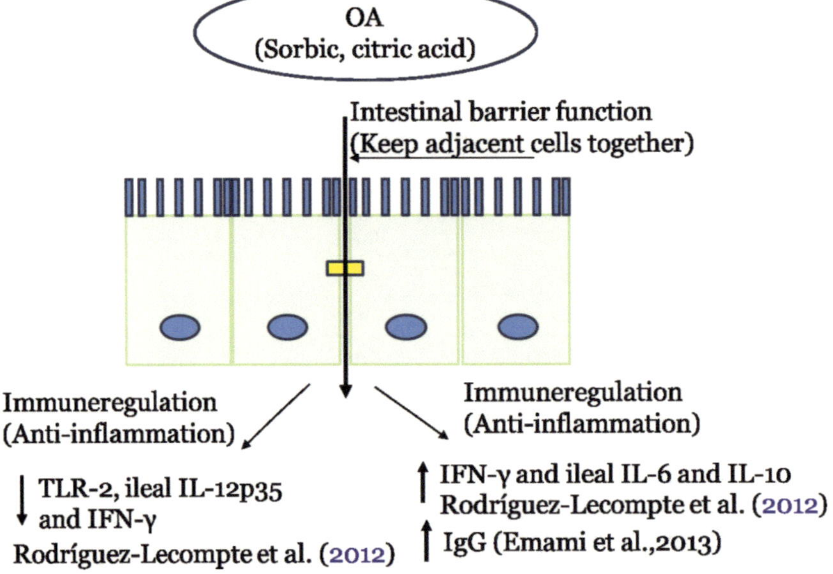

Figure 1.
The role of organic acids (sorbic and acid citric) in the intestinal anti-inflammation and immune response in broiler chickens.

several mechanisms, including the reduction of pro-inflammatory cytokines (INF-γ, TNF-α, IL-1β, IL-6, and IL-8), while also including IL-10 and TGF- β (**Figure 1**).

With this property, butyrate enhances intestinal barrier function and mucosal immunity leading to the enhanced protection against luminal pathogens [52]. For example, feeding the ApoE knockout mice with butyrate decreased the pro-inflammatory cytokines, leading to a reduction in atherosclerotic lesions and a decrease in macrophage migration [54]. Kim et al. (2013) found that SCFAs activate GPR41 and GPR43 in mice intestinal epithelial cells, leading to the production of chemokines and cytokines, which are required for an inflammatory response to bacterial infection [55]. Rodríguez-Lecompte et al. (2012) indicated that broiler chicks fed with probiotics (*Lactobacillus casei*, *Lactobacillus acidophillus*, *Streptococcus faecium*, and *Saccharomyces cerevisiae*) and organic acids (sorbic and citric acid) positively responded to anti-inflammatory via pathways involving cytokines by decreasing TLR-2 and ileal IL-12p35 and increasing IFN-γ and ileal IL-6 and IL-10 [56]. In addition, IgA (SIgA) is the most prominent antibody produced in the intestinal mucosa that protects the intestines against bacterial and viral infections [51]. Schilderink et al. reported that acetate increased fecal IgA and IgA-positive B-cells in the lamina propria of wild-type mice indicating that the process was mediated through specific SCFA receptor interaction [57]. Emami et al. found that broilers fed with phytase and organic acids showed higher IgG in the primary and secondary response compared to the control group [58]. Park et al. noticed that the supplementation of 0.2% organic acid to layer diet aged 75 weeks significantly increased IgY level [59].

4. Effect of acidifiers on swine and poultry production

4.1 Effect of acidifiers on swine production

Previous research showed positive effects of supplementing dietary acidifiers at optimal levels on the performance and gut health of swine at different growth stages (**Table 3**). For example, Li et al. (2008) reported that weanling piglets fed a diet supplemented with 0.5% of a mixture of acidifiers, including calcium salt of 2-hydrozy-4(methylthio) butanoic acid, fumaric acid, and benzoic acid) exhibited better weight gain and feed efficiency ($p < 0,05$), higher levels of lactobacilli in the duodenum, and lower levels of ileal *E. coli* [71]. Kuang et al. (2015) also noted that 21-day-old crossbred pigs, when fed a diet supplemented with 0.3% blends of acidifiers containing citric acid, calcium formate, calcium lactate, and MCFAs (capric, lauric, and myristic acids), experienced improvements in ADG, average daily feed intake (ADFI), increased AA digestibility, and enhanced immunity [72].

It is reported that supplementation of 0.4% acidifier mixture (fumaric, lactic, propionic acids, citric, benzoic) in the dietary of weaning piglets improved the growth performance, feed intake (FI) and gain-to-feed ratio (G: F) compared to the diet without acidifiers supplementation [73]. Regarding the growing pigs and finishing pigs, it is also demonstrated that the supplementation of 0.2% of coated organic acids in the dietary including 10% malic, 13% citric, 17% fumaric acids, and 1.2% MCFA (capric and caprylic acid) has a positive influence on the growth performance. Feces from pigs fed a diet supplemented with this organic acid blend showed a linear reduction ($p < 0.001$) in *E. coli* counts and a tendency for a linear increase ($p = 0.06$) in *Lactobacillus* counts [74]. Zhai et al. (2017) reported that the nursery and grower-finisher pigs fed with the supplementation levels of 0.3 and 0.5% benzoic

Composition of acidifiers	Dose	Age	Growth performance			Gut health	Ref
			ADG	ADF1	G:F		
Single acidifiers							
Fumaric	0,15%, 0,3%	Weaned	*	*	*	NA	[60]
Benzoic	0,3%; 0,5%	Nursery, Grower, Finisher	*	*	*	NA	[61]
Lactic	2,8%	Weaned	NA	NA	NA	Control clinical and subclinical infections of *S. Typhimurium*	[62]
	1,6%		*	*	*	Reduced incidence and severity of diarrhea	[63]
Formic	1,2%	Weaned	*	*	*	Reduced incidence and severity of diarrhea	[63]
Propionic	1,0%	Weaned	*	*	*	Reduced incidence and severity of diarrhea	[63]
Citric acid	1,0%	Weaned	NS	NS	*	Improved intestinal morphology	[26]
Mixture of acidifier							
Formic acid, acetic acid, propionic acid, and butyric acid	1,5 g/kg	Weaned	*	*	*	Increased *lactobacillus*,	[64]
Formic acid, acetic acid, and propionic acid, medium-chain fatty acids (MCFA)		Weaned	*	*	*	Improved intestinal structure	[65]
Formic acid (31.0%), ammonium formate (23.0%), and acetic acid (8.3%)	2 L/ton in drinking water	Weaned	NS	*	NS	Decreased diarrhea rate, regulate gut microbiota	[66]
Formic acid (11%), ammonium formate (13%), propionic acid (10%), acetic acid (5.1%), and citric acid (3.7%)	3 g/kg 5 g/kg	Weaned	*	NS	*	Improved intestinal morphology	[67]
Salts of acidifier							
Encapsulated sodium butyrate	30.00%	Growing-finishing	*	NS	NS	NA	[68]
Sodium butyrate	0.8 g/kg	Weaned	*	*	*	NA	[69]

Composition of acidifiers	Dose	Age	Growth performance			Gut health	Ref
			ADG	ADF1	G:F		
Coated sodium butyrate	300 mg/kg 450 mg/kg	Weaned	*	*	*	*Increased lactobacillus*, decreased *E. coli* counts	[70]

*NA: not available, NS: not significant difference in p-value, ADFI: average daily feed intake, ADG: average daily gain, G:F: gain: feed, *: significant effect of OAs on growth performance (p < 0,05).*

Table 3.
Effects of acidifiers on growth performance and gut health of swine.

acid showed a significant improvement in growth performance. In which, the supplementation of 0.5% benzoic acid promoted better performance in nursery pigs, while grower-finisher pigs fed with 0.36% gained optimal ADG [61].

Moreover, evidence also showed the importance of organic acids on gut health and livestock environment. For example, addition of benzoic acid (1 or 2%) in the dietary for grower-finisher pigs reduced urinary pH and NH3 emissions [75, 76]. Diao et al. (2014) also reported that benzoic acid supplementation (5 g/kg) in the dietary decreased the GIT pH values. The number of *Bifidobacterium* and *Bacillus* in pigs fed the benzoic acid diet was greater than in pigs fed the control diet, while the number of *Escherichia coli* decreased in pigs fed the benzoic acid diet. In addition, benzoic acid increased the content of propionic acid and total volatile fatty acids and decreased the concentrations of NH_3–N in cecum ($P < 0.05$). The gut morphology was also improved in pigs fed the benzoic acid diet ($P < 0.05$), with observed increases in villus height in the ileum and decreased crypt depth in the duodenum [77]. Lynch et al. (2017) indicated a significant decrease in *Salmonella* levels in the feces of grower pigs fed with sodium butyrate (p = 0.001) and a blend of formic and citric acids (p < 0.001) [78]. Zhang et al. (2018) showed that dietary supplementation with chlorogenic acid improved intestinal health and regulated the composition of selected intestinal microbiota in weaned piglets. To put it more specific, an increase in the population of *Lactobacillus* (p < 0.05) and a decrease in the population of *E. coli* were observed in the colon of pigs fed chlorogenic acid diets. Dietary supplementation with chlorogenic acid also resulted in an increase (p < 0.05) in duodenal villus height and villus height: crypt depth compared to the control group. This positive influence on intestinal morphology in weaned piglets ultimately improved their growth performance [79].

In addition, the recent study showed the effect of a microencapsulated mixture of organic acids (MOAs) supplementation on the growth performance and meat-carcass grade quality in growing-finishing pigs. The supplementation of MOAs (0,05 and 0,1%) in the basal diet resulted in a significant ($P < 0.05$) linear improvement in ADG, a linear decrease in fecal *E. coli* counts, a linear ($P < 0.05$) increase in backfat thickness and lean meat percentage, and a decrease in drip loss [33]. Similarly, the previous trial showed that piglets received a basal diet with the addition of MOAs at 3 kg/ton had higher ADFI (+ 4.6%; $P = 0.08$), ADG (+ 8%; $P < 0.01$), and final body weight (+ 6.5%, $P < 0.01$) [80]. Nguyen et al. indicated that the administration of MOAs (0,1 and 0,2% in the diets) *increased Lactobacillus* counts and decreased *E. coli* counts compared to the control diet (p < 0.05) [62]. These findings suggest that organic acids have growth-promoting properties and can be used as alternatives to antibiotics in swine production.

4.2 Effect of acidifiers on poultry production

Acidifiers and their salts have also been used in poultry dietary and drinking water for the past decades. Literature showed that the broilers/layers fed with acidifiers in the diet improved growth performance, reduced toxic bacterial mass, and enhanced nutrient digestibility and GIT immunity (**Table 4**).

When it comes to broiler growth performance, previous trials have demonstrated the efficiency of supplementing diets with butyric acid and its salt (sodium butyrate) in improving body weight, feed intake, and FCR. For instance, Leeson et al. (2005) and Anton Giovanni et al. (2007) showed that the carcass weight and breast meat yield significantly increased ($p < 0.05$) in birds fed 0.2% butyric acid [91, 92]. Besides, Adil et al. (2011) found that birds fed 3% fumaric acid exhibited significantly ($p < 0.05$) higher body weight gains and better feed conversion ratio [93].

For the combination of organic acids, Nguyen et al. (2018) reported that broilers fed with various levels of mixed acidifiers (0.02, 0.03, 0.04, 0.05, and 0.06%) and MCFAs showed positive growth performance, nutrient digestibility, and excreta microflora. In detail, broilers exhibited a linear increase ($P < 0.05$) in body weight gain and an improvement in feed conversion ratio ($P < 0.0001$). Additionally, there was a linear increase ($P < 0.05$) in the *Lactobacillus*: *E. coli* ratio. An increase in the levels of organic acids and MCFAs also significantly improved the IgG concentration ($P = 0.011$) [86]. However, Youshelf et al. (2017) reported that supplements of single lactic acid (0,2%) in broiler diets seem to obtain better performances than the organic acid mixture (0,4%). It was also found that the inclusion of single lactic acid in broiler diets declined the serum cholesterol level, the pH of small intestine, the counts of fecal coliforms and *E. coli*, but did not affect the carcass yield, breast, or organ weights [94].

In addition, salts of organic acids, such as potassium diformate and sodium diformate have been shown to have positive effects on performance and GIT health. To put it more specific, Paul et al. (2007) reported that ammonium formate or calcium propionate (0.3%) increased the live weight gain and FCR at day 21 in broiler chickens [95]. Mikkelsen et al. (2009) showed that inclusion of 0.45% potassium diformate reduced mortality caused by necrotic enteritis (*Clostridium perfringens*) [96]. Raaga et al. (2016) reported that broilers fed basal diet supplemented with formic acid (5 g/kg diet), or potassium diformate (5 g/kg diet) exhibited significantly increased body weight gain and improved feed conversion ratio ($P < 0.05$). An improvement in villus height was also observed in both of these groups. [97]. Besides, different organic acids have been used in drinking water. Formic, propionic acids, and their salts have exceptionally good solubility in water. Their supplementation in drinking water with 0,3 L/1000 L significantly improved the intestinal structure [98].

In the laying hen industry, the efficiency of dietary acidifiers on egg production and quality have been well-documented. Yesilbag and Çolpan (2006) reported that the laying hens fed with a mixture of acidifiers at levels of 0,5%, 1,0%, and 1,5% exhibited a slight increase in average egg production (91.03, 90.94, and 91.30%, respectively) compared to the control group (85.76%) [99]. Grashorn et al. (2013) showed that the supplementation of organic acids mixture (SALMO-NIL dry) at 2 kg/ton of feed increased average egg weight and egg production capacity [100]. Recently, Gong et al. (2021) reported that the dietary supplementation with 1 g/kg benzoic acid exhibited no effect on production performance, but it significantly improved egg quality, intestinal morphology, and bacterial profiles [101]. Encapsulation technology is also currently employed in laying hen industry to produce protected organic acids.

Composition of acidifiers	Dose	Age	Growth performance			Gut health	Ref
			ADG	ADF1	G: F		
Single acidifiers							
Phosphoric	0.1%, 0.2%	1–42 days old	*	*	*	Decreased *E. coli, Salmonella*	[81]
Lactic	0.3%	1–42 days old	*	*	*	Decreased *E. coli, Salmonella*	[81]
Propionic	0.5%	1–42 days old	*	*	*	Increased *Lactobacillus*, decreased *E. coli*	[82]
Formic	0.5%	1–42 days old	*	*	*	Increased *Lactobacillus*, decreased *E. coli*	[82]
Formic	0.4%	1–48 days old	*	*	*	NA	[83]
Citric	0.3%	1–42 days old	*	*	*	Improved gut morphology	[84]
Encapsulated Butyric	0.03%; 0.05%	1–42 days old	*	*	*	NA	[85]
Mixture of acidifier							
17% fumaric acid, 13% citric acid, 10% malic acid, and 1.2% MCFAs	0.06%	Broiler	*	*	*	Increased IgG, increased *Lactobacillus*, decreased *E. coli*	[86]
Formic, propionic	0,2%; 0,4%	Starter, Grower, Finisher broiler	*	*	*	Increased *Lactobacillus*, decreased *E. coli*	[87]
Formic acid 31%, propionic acid 19%, ammonium format 26%, ammonium propionate 6%	0,3 L/1000 L drinking water	1–42 days old	*	*	*	Improved intestinal structure	[88]
Salts of acidifiers							
Sodium butyrate	500, 1000, 2000 mg/kg	1–42 days old	*	*	*	Improved intestinal structure, increased *Lactobacillus*	[89]

Composition of acidifiers	Dose	Age	Growth performance			Gut health	Ref
			ADG	ADF1	G: F		
Sodium butyrate	0.3 g/kg; 0.6 g/kg; 1.2 g/kg	1–21 days old	*	*	*	Improved intestinal structure, enhanced the immune response of ND vaccine.	[90]

*NA: not available, NS: not significant difference in p-value, ADFI: average daily feed intake, ADG: average daily gain, G:F: gain: feed, *: significant effect of OAs on growth performance (p < 0,05).*

Table 4.
Effects of acidifiers on growth performance and gut health of broilers.

Youself et al. (2013) evaluated the effect of microencapsulated organic acids including fumaric acid, calcium formate, calcium propionate, potassium sorbate on egg quality. The results showed that microencapsulated organic acids did not affect shape index, yolk index, Haugh unit or specific gravity, but showed significant increase in shell thickness and yolk color [102]. Recently, Garcia et al. (2019) showed the effects of beak trimming and the inclusion of sodium butyrate in the diet from at hatch to 6 weeks of age on the growth performance and GIT traits of brown-egg pullets. The results showed that sodium butyrate tended to improve growth and FCR from 0 to 6 weeks of age but did not affect body weight uniformity [103].

In addition, drinking water acidification is also preferred in layer industry for improving performance. Kadim et al. (2008) reported that the average egg production significantly increased by approximately 20, 15, and 10% in the trial groups where acetic acid was administered through drinking water at levels of 0.06, 0.04, and 0.02%, respectively, during the hot season (P < .01) [104]. Abbas et al. (2013) indicated that administration of formic acid through drinking water at levels of 0, 0.05, 0.10, or 0.15% increased average egg production in hens by approximately 72, 80, 86, and 88%, respectively [105].

5. Conclusions

From the scientific results presented and discussed in this chapter, the following main conclusions can be drawn: (i) OAs and their salts are among the most promising future products of the livestock industry, owing to their antimicrobial activity, which reflect in improved overall gut health, inhibition of pathogenic bacteria growth, increased apparent total tract digestibility, and enhanced growth performance (ii) Both single OAs and mixed OAs are utilized as additives in swine and poultry feeds, and have positive influences on growth performance and gut health in the different growth periods of swine and poultry. In which, the mixed OAs seem to be more favorable for recent investigations shown with the enormous number of publications (iii) the different forms of OAs such as their salts and derivates seem to be more efficacy for the growth performance and gut health of pig and poultry compared to original OA forms. (iv) OAs can be added in drinking water or in the dietary of swine and poultry. Both supplementation methods were evaluated to improve the growth performance and control pathogenic bacteria.

Author details

Nguyen Vu Thuy Hong Loan[1], Ho Trung Thong[1*], Le Nu Anh Thu[2] and Ho Viet Duc[3]

1 Faculty of Veterinary Medicine and Animal Husbandry, HUTECH University, Ho Chi Minh City, Vietnam

2 Faculty of Animal Science and Veterinary Medicine, University of Agriculture and Forestry, Hue University, Hue City, Vietnam

3 School of Engineering, The Hong Kong University of Science and Technology, Hong Kong, China

*Address all correspondence to: hothonghuaf@gmail.com

References

[1] Castanon JIR. History of the use of antibiotic as growth promoters in European poultry feeds. Poultry Science. 2007;**86**:2466-2471. DOI: 10.3382/ps.2007-00249

[2] Jones FT, Ricke SC. Observations on the history of the development of antimicrobials and their use in poultry feeds. Poultry Science. 2003;**82**:613-617. DOI: 10.1093/ps/82.4.613

[3] Brown K, Uwiera RRE, Kalmokoff ML, Brooks SPJ, Inglis GD. Antimicrobial growth promoter use in livestock: A requirement to understand their modes of action to develop effective alternatives. International Journal of Antimicrobial Agents. 2017;**49**:12-24. DOI: 10.1016/j.ijantimicag.2016.08.006

[4] Khan SH, Iqbal J. Recent advances in the role of organic acids in poultry nutrition. Journal of Applied Animal Research. 2016;**44**(1):359-369. DOI: 10.1080/09712119.2015.1079527

[5] Laxminarayan R, Van Boeckel T, Teillant A. The economic costs of withdrawing antimicrobial growth promoters from the livestock sector. OECD Food, Agriculture and Fisheries Papers 78. Paris: OCED Publishing; 2015. DOI: 10.1787/5js64kst5wvl-en.7

[6] Adil S, Banday T, Bhat GA, Mir MS, Rehman M. Effect of dietary supplementation of organic acids on performance, intestinal histomorphology, and serum biochemistry of broiler chicken. Veterinary Medicine International. 2010;**2010**. DOI: 10.4061/2010/479485

[7] Salaheen S, Kim SW, Haley BJ, Van Kessel JAS, Biswas D. Alternative growth promoters modulate broiler gut microbiome and enhance body weight gain. Frontiers in Microbiology. 2017;**26**:8. DOI: 10.3389/fmicb.2017.02088

[8] Gernat AA, Santos FBO, Grimes JL. Alternative approaches to antimicrobial use in the Turkey industry: Challenges and perspectives. German Journal of Veterinary Research. 2021;**1**(3):37-47. DOI: 10.51585/gjvr.2021.3.0018

[9] Lillehoj H, Liu Y, Calsamiglia S, Fernandez-Miyakawa ME, Chi F, Cravens RL, et al. Phytochemicals as antibiotic alternatives to promote growth and enhance host health. Veterinary Research. 2018;**49**(1):76. DOI: 10.1186/s13567-018-0562-6

[10] Liao SF, Nyachoti M. Using probiotics to improve swine gut health and nutrient utilization. Animal Nutrition. 2017;**3**:331-343. DOI: 10.1016/j.aninu.2017.06.007

[11] Salim HM, Huque KS, Kamaruddin KM, Beg MAH. Global restriction of using antibiotic growth promoters and alternative strategies in poultry production. Science Progress. 2018;**101**(1):52-75. DOI: 10.3184/003685018X15173975498947

[12] Gadde U, Kim WH, Oh ST, Lillehoj HS. Alternatives to antibiotics for maximizing growth performance and feed efficiency in poultry: A review. Animal Health Research Reviews. 2017;**18**(1):26-45. DOI: 10.1017/S1466252316000207

[13] Nusairat B, Wang JJ. Xylanase and direct-fed microbials (DFM) potential for improvement of live performance, energy digestibility, and reduction of environmental microbial load

of broilers. Frontiers in Veterinary Science. 2020;7:606415. DOI: 10.3389/fvets.2020.606415

[14] Wang S, Zeng X, Yang Q, Qiao S. Antimicrobial peptides as potential alternatives to antibiotics in food animal industry. International Journal of Molecular Sciences. 2016;**17**(5):603. DOI: 10.3390/ijms17050603

[15] Eid S, Tolba HMN, Hamed RI, Al-Atfeehy NM. Bacteriophage therapy as an alternative biocontrol against emerging multidrug resistant E. coli in broilers. Saudi Journal of Biological Sciences. 2022;**29**(5):3380-3389. DOI: 10.1016/j.sjbs.2022.02.015

[16] Burel C. Alternatives to antimicrobial growth promoters (AGPs) in animal feed. In: Animal Feed Contamination: Effects on Livestock and Food Safety. Cambridge: Woodhead Publishing, Elsevier Ltd.; 2012. pp. 432-448. DOI: 10.1533/9780857093615.4.432

[17] Pearlin BV, Muthuvel S, Govidasamy P, Villavan M, Alagawany M, Ragab Farag M, et al. Role of acidifiers in livestock nutrition and health: A review. Journal of Animal Physiology and Animal Nutrition. 2020;**104**:558-569. DOI: 10.1111/jpn.13282

[18] Khan RU, Naz S, Raziq F, Qudratullah Q, Khan NA, Laudadio V, et al. Prospects of organic acids as safe alternative to antibiotics in broiler chickens diet. Environmental Science and Pollution Research. 2022;**29**(22):32594-32604. DOI: 10.1007/s11356-022-19241-8

[19] Suiryanrayna MVAN, Ramana JV. A review of the effects of dietary organic acids fed to swine. Journal of Animal Science and Biotechnology. 2015;**6**(1):45. Available from: http://jasbsci.biomedcentral.com/articles/10.1186/s40104-015-0042-z

[20] Nguyen DH, Seok WJ, Kim IH. Organic acids mixture as a dietary additive for pigs—a review. Animals. 2020;**10**(6):952. DOI: 10.3390/ani10060952

[21] Grilli E, Piva A. Organic acids and their role in reduce foodborne pathogens in food animals. In: Callaway TR, Edrington TS, editors. On-Farm Strategies to Control Foodborne Pathogens. Hauppauge, NY, USA: Nova Science Pub. Inc; 2012. pp. 183-210

[22] Liu L, Li Q, Yang Y, Guo A. Biological function of short-chain fatty acids and its regulation on intestinal health of poultry. Frontiers in Veterinary Science. 2021;**8**:736739. DOI: 10.3389/fvets.2021.736739

[23] Tugnoli B, Giovagnoni G, Piva A, Grilli E. From acidifiers to intestinal health enhancers: How organic acids can improve growth efficiency of pigs. Animals. 2020;**10**(1):134. DOI: 10.3390/ani10010134

[24] Jackman JA, Boyd RD, Elrod CC. Medium-chain fatty acids and monoglycerides as feed additives for pig production: Towards gut health improvement and feed pathogen mitigation. Journal of Animal Science and Biotechnology. 2020;**11**(1):44. DOI: 10.1186/s40104-020-00446-1

[25] Melaku M, Zhong R, Han H, Wan F, Yi B, Zhang H. Butyric and citric acids and their salts in poultry nutrition: Effects on gut health and intestinal microbiota. International Journal of Molecular Sciences. 2021;**22**(19):10392. DOI: 10.3390/ijms221910392

[26] Ferreira JL, Watanabe PH, Mendonça IB, Nogueira BD, Ferreira ACS, Nepomuceno RC, et al. Calcium anacardate and citric acid as growth promoters for weaned piglets.

Livestock Science. 2020;**238**:104084. DOI: 10.1016/j.livsci.2020.104084

[27] Ferronato G, Prandini A. Dietary supplementation of inorganic, organic, and fatty acids in pig: A review. Animals. 2020;**10**(10):1740. DOI: 10.3390/ani10101740

[28] Dittoe DK, Ricke SC, Kiess AS. Organic acids and potential for modifying the avian gastrointestinal tract and reducing pathogens and disease. Frontiers in Veterinary Science. 2018;**5**:216. DOI: 10.3389/fvets.2018.00216

[29] Bedford A, Gong J. Implications of butyrate and its derivatives for gut health and animal production. Animal Nutrition. 2018;**4**(2):151-159. DOI: 10.1016/j.aninu.2017.08.010

[30] Onrust L, Van Driessche K, Ducatelle R, Schwarzer K, Haesebrouck F, Van Immerseel F. Valeric acid glyceride esters in feed promote broiler performance and reduce the incidence of necrotic enteritis. Poultry Science. 2018;**97**(7):2303-2311. Available from: https://www.sciencedirect.com/science/article/pii/S0032579119303438

[31] Öztürk E, Temiz U. Encapsulation methods and use in animal nutrition. Selcuk Journal of Agricultural and Food Sciences. 2018;**32**(3):624-631. DOI: 10.15316/SJAFS.2018.145

[32] Rajendran D, Ezhuthupurakkal PB, Lakshman R, Gowda NKS, Manimaran A, Rao SBN. Application of encapsulated nano materials as feed additive in livestock and poultry: A review. Veterinary Research Communications. 2022;**46**(2):315-328. DOI: 10.1007/s11259-022-09895-7

[33] Muniyappan M, Palanisamy T, Kim IH. Effect of microencapsulated organic acids on growth performance, nutrient digestibility, blood profile, fecal gas emission, fecal microbial, and meat-carcass grade quality of growing-finishing pigs. Livestock Science. 2021;**252**:104658. Available from: https://www.sciencedirect.com/science/article/pii/S1871141321002663

[34] Feye KM, Swaggerty CL, Kogut MH, Ricke SC, Piva A, Grilli E. The biological effects of microencapsulated organic acids and botanicals induces tissue-specific and dose-dependent changes to the Gallus gallus microbiota. BMC Microbiology. 2020;**20**(1):332. DOI: 10.1186/s12866-020-02001-4

[35] Hajati H. Application of organic acids in poultry nutrition. International Journal of Avian & Wildlife Biology. 2018;**3**(4):324-329. DOI: 10.15406/ijawb.2018.03.00114

[36] Papatsiros G, Billinis C. The prophylactic use of acidifiers as antibacterial agents in swine. In: Antimicrobial Agents. London, UK: InTech; 2012. DOI: 10.5772/32278

[37] Ao T, Cantor AH, Pescatore AJ, Ford MJ, Pierce JL, Dawson KA. Effect of enzyme supplementation and acidification of diets on nutrient digestibility and growth performance of broiler chicks1. Poultry Science. 2009;**88**(1):111-117. Available from: https://www.sciencedirect.com/science/article/pii/S0032579119389035

[38] Ghazalah AA, Atta AM, Elkloub MK, MEL M, Shata R. Effect of dietary supplementation of organic acids on performance, nutrients digestibility and health of broiler chicks. International Journal of Poultry Science. 2011;**10**:176-184. DOI: 10.3923/ijps.2011.176.184

[39] García V, Catalá-Gregori P, Hernández F, Megías MD, Madrid J.

Effect of formic acid and plant extracts on growth, nutrient digestibility, intestine mucosa morphology, and meat yield of broilers. Journal of Applied Poultry Research. 2007;**16**(4):555-562. Available from: https://www.sciencedirect.com/science/article/pii/S1056617119316320

[40] Upadhaya S, Lee K, Kim I. Protected organic acid blends as an alternative to antibiotics in finishing pigs. Asian-Australasian Journal of Animal Sciences. 2014;**27**:1600-1607. DOI: 10.5713/ajas.2014.14356

[41] Panda AK, Rao SVR, Raju MVLN, Sunder GS. Effect of butyric acid on performance, gastrointestinal tract health and carcass characteristics in broiler chickens. Asian-Australasian Journal of Animal Sciences. 2009;**22**(7):1026-1031. DOI: 10.5713/ajas.2009.80298

[42] Gálfi P, Bokori J. Feeding trial in pigs with a diet containing sodium n-butyrate. Acta Veterinaria Hungarica. 1990;**38**(1-2):3-17. Available from: http://europepmc.org/abstract/MED/2100936

[43] Panah FM, Lauridsen C, Højberg O, Jensen HE, Nielsen TS. Composition of mucus- and digesta-associated bacteria in growing pigs with and without diarrhea differed according to the presence of colonic inflammation. BMC Microbiology. 2023;**23**(1):145. Available from: https://bmcmicrobiol.biomedcentral.com/articles/10.1186/s12866-023-02874-1

[44] Argüello H, Estellé J, Zaldívar-López S, Jiménez-Marín Á, Carvajal A, López-Bascón MA, et al. Early Salmonella Typhimurium infection in pigs disrupts microbiome composition and functionality principally at the ileum mucosa. Scientific Reports. 2018;**8**(1):7788. DOI: 10.1038/s41598-018-26083-3

[45] Immerseel F, Russell J, Flythe M, Gantois I, Timbermont L, Pasmans F, et al. The use of organic acids to combat Salmonella in poultry: A mechanistic explanation of the efficacy. Avian Pathology. 2006;**35**:182-188. DOI: 10.1080/03079450600711045

[46] Hsiao CP, Siebert KJ. Modeling the inhibitory effects of organic acids on bacteria. International Journal of Food Microbiology. 1999;**47**(3):189-201. DOI: 10.1016/S0168-1605(99)00012-4

[47] Peh E, Kittler S, Reich F, Kehrenberg C. Antimicrobial activity of organic acids against Campylobacter spp. and development of combinations-a synergistic effect? PLoS One. 2020;**15**(9):e0239312. DOI: 10.1371/journal.pone.0239312

[48] Huyghebaert G, Ducatelle R, Van Immerseel F. An update on alternatives to antimicrobial growth promoters for broilers. The Veterinary Journal. 2011;**187**(2):182. Available from: https://www.sciencedirect.com/science/article/pii/S1090023310000869

[49] Kirchgessner M, Roth FX. Fumaric acid as a feed additive in pig nutrition. Pig News and Information. 1982;**3**:259-264

[50] Blank R, Mosenthin R, Sauer WC, Huang S. Effect of fumaric acid and dietary buffering capacity on ileal and fecal amino acid digestibilities in early-weaned pigs. Journal of Animal Science. 1999;**77**(11):2974-2984. DOI: 10.2527/1999.77112974x

[51] Blaak EE, Canfora EE, Theis S, Frost G, Groen AK, Mithieux G, et al. Short chain fatty acids in human gut and metabolic health. Beneficial Microbes. 2020;**11**(5):411-455. DOI: 10.3920/BM2020.0057

[52] Liu H, Wang J, He T, Becker S, Zhang G, Li D, et al. Butyrate: A double-edged sword for health? Advances in Nutrition. 2018;**9**(1):21-29. DOI: 10.1093/advances/nmx009

[53] Donohoe DR, Garge N, Zhang X, Sun W, O'Connell TM, Bunger MK, et al. The microbiome and butyrate regulate energy metabolism and autophagy in the mammalian colon. Cell Metabolism. 2011;**13**(5):517-526. Available from: https://www.sciencedirect.com/science/article/pii/S1550413111001434

[54] Aguilar EC, Leonel AJ, Teixeira LG, Silva AR, Silva JF, Pelaez JMN, et al. Butyrate impairs atherogenesis by reducing plaque inflammation and vulnerability and decreasing NFkB activation. Nutrition, Metabolism and Cardiovascular Diseases. 2014;**24**(6):606-613. DOI: 10.1016/j.numecd.2014.01.002

[55] Kim MH, Kang SG, Park JH, Yanagisawa M, Kim CH. Short-chain fatty acids activate GPR41 and GPR43 on intestinal epithelial cells to promote inflammatory responses in mice. Gastroenterology. 2013;**145**(2):396-406. DOI: 10.1053/j.gastro.2013.04.056

[56] Rodríguez-Lecompte JC, Yitbarek A, Brady J, Sharif S, Cavanagh MD, Crow G, et al. The effect of microbial-nutrient interaction on the immune system of young chicks after early probiotic and organic acid administration. Journal of Animal Science. 2012;**90**(7):2246-2254

[57] Schilderink R, Verseijden C, Seppen J, Muncan V, van den Brink GR, Lambers TT, et al. The SCFA butyrate stimulates the epithelial production of retinoic acid via inhibition of epithelial HDAC. American Journal of Physiology-Gastrointestinal and Liver Physiology. 2016;**310**(11):G1138-G1146. DOI: 10.1152/ajpgi.00411.2015

[58] Khodambashi Emami N, Zafari Naeini S, Ruiz-Feria CA. Growth performance, digestibility, immune response and intestinal morphology of male broilers fed phosphorus deficient diets supplemented with microbial phytase and organic acids. Livestock Science. 2013;**157**(2):506-513. Available from: https://www.sciencedirect.com/science/article/pii/S1871141313003648

[59] Park KW, Rhee AR, Um JS, Paik IK. Effect of dietary available phosphorus and organic acids on the performance and egg quality of laying hens. Journal of Applied Poultry Research. 2009;**18**(3):598-604. DOI: 10.3382/japr.2009-00043

[60] Radecki SV, Juhl MR, Miller ER. Fumaric and citric acids as feed additives in starter pig diets: Effect on performance and nutrient balance. Journal of Animal Science. 1988;**66**(10):2598-2605. DOI: 10.2527/jas1988.66102598x

[61] Zhai H, Ren W, Wang S, Wu J, Guggenbuhl P, Kluenter AM. Growth performance of nursery and grower-finisher pigs fed diets supplemented with benzoic acid. Animal Nutrition. 2017;**3**(3):232-235. DOI: 10.1016/j.aninu.2017.05.001

[62] Nguyen DH, Lee KY, Tran HN, Kim IH. Effect of a protected blend of organic acids and medium-chain fatty acids on growth performance, nutrient digestibility, blood profiles, meat quality, faecal microflora, and faecal gas emission in finishing pigs. Canadian Journal of Animal Science. 2018;**99**(3):448-455. DOI: 10.1139/cjas-2016-0174

[63] Tsiloyiannis VK, Kyriakis SC, Vlemmas J, Sarris K. The effect of organic acids on the control of porcine post-weaning diarrhoea. Research in Veterinary Science. 2001;**70**(3):287-293. Available from: https://www.

sciencedirect.com/science/article/pii/S003452880190476X

[64] Xu YT, Li L, Long SF, Pan L, Piao XS. Effect of organic acids and essential oils on performance, intestinal health and digestive enzyme activities of weaned pigs. Animal Feed Science and Technology. 2018;**235**:110-119. Available from: https://www.sciencedirect.com/science/article/pii/S0377840117309690

[65] Long SF, Xu YT, Pan L, Wang QQ, Wang CL, Wu JY, et al. Mixed organic acids as antibiotic substitutes improve performance, serum immunity, intestinal morphology and microbiota for weaned piglets. Animal Feed Science and Technology. 2018;**235**:23-32. Available from: https://www.sciencedirect.com/science/article/pii/S0377840116311671

[66] Xiang XD, Deng ZC, Wang YW, Sun H, Wang L, Han YM, et al. Organic acids improve growth performance with potential regulation of redox homeostasis, immunity, and microflora in intestines of weaned piglets. Antioxidants. 2021;**10**(11):1665. DOI: 10.3390/antiox10111665

[67] Ma J, Piao X, Shang Q, Long S, Liu S, Mahfuz S. Mixed organic acids as an alternative to antibiotics improve serum biochemical parameters and intestinal health of weaned piglets. Animal Nutrition. 2021;**7**(3):737-749. DOI: 10.1016/j.aninu.2020.11.018

[68] da Silva CA, Bridi AM, Dias CP, Callegari MA, Nunes EC, Pierozan CR, et al. Effects of encapsulated sodium butyrate and phytogenic on growth performance, carcass traits and health of growing-finishing pigs. Ciencia Rural. 2020;**50**(11):1-10. DOI: 10.1590/0103-8478cr20190718

[69] Piva A, Morlacchini M, Casadei G, Gatta PP, Biagi G, Prandini A. Sodium butyrate improves growth performance of weaned piglets during the first period after weaning. Italian Journal of Animal Science. 2002;**1**(1):35-41. DOI: 10.4081/ijas.2002.35

[70] Lin F, Li X, Wen J, Wang C, Peng Y, Feng J, et al. Effects of coated sodium butyrate on performance, diarrhea, intestinal microflora and barrier function of pigs during the first 2-week post-weaning. Animal Feed Science and Technology. 2020;**263**:114464. Available from: https://www.sciencedirect.com/science/article/pii/S0377840118311581

[71] Li Z, Yi G, Yin J, Sun P, Li D, Knight C. Effects of organic acids on growth performance, gastrointestinal pH, intestinal microbial populations and immune responses of weaned pigs. Asian-Australasian Journal of Animal Sciences. 2008;**21**(2):252-261. DOI: 10.5713/ajas.2008.70089

[72] Kuang Y, Wang Y, Zhang Y, Song Y, Zhang X, Lin Y, et al. Effects of dietary combinations of organic acids and medium chain fatty acids as a replacement of zinc oxide on growth, digestibility and immunity of weaned pigs. Animal Feed Science and Technology. 2015;**208**:145-157. Available from: https://www.sciencedirect.com/science/article/pii/S0377840115002382

[73] Walsh MC, Sholly DM, Hinson RB, Saddoris KL, Sutton AL, Radcliffe JS, et al. Effects of water and diet acidification with and without antibiotics on weanling pig growth and microbial shedding. Journal of Animal Science. 2007;**85**(7):1799-1808. DOI: 10.2527/jas.2006-049

[74] Upadhaya SD, Lee KY, Kim IH. Effect of protected organic acid blends on growth performance, nutrient digestibility and faecal micro flora in growing pigs. Journal of Applied

Animal Research. 2016;**44**(1):238-242. DOI: 10.1080/09712119.2015.1031775

[75] Murphy DP, O'Doherty JV, Boland TM, O'Shea CJ, Callan JJ, Pierce KM, et al. The effect of benzoic acid concentration on nitrogen metabolism, manure ammonia and odour emissions in finishing pigs. Animal Feed Science and Technology. 2011;**163**(2):194-199. Available from: https://www.sciencedirect.com/science/article/pii/S0377840110003597

[76] Nørgaard JV, Fernández JA, Sørensen KU, Wamberg S, Poulsen HD, Kristensen NB. Effect of benzoic acid supplementation on acid–base status and mineral metabolism in catheterized growing pigs. Livestock Science. 2010;**134**(1):116-118. Available from: https://www.sciencedirect.com/science/article/pii/S1871141310003276

[77] Diao H, Zheng P, Yu B, He J, Mao XB, Yu J, et al. Effects of dietary supplementation with benzoic acid on intestinal morphological structure and microflora in weaned piglets. Livestock Science. 2014;**167**:249-256. Available from: https://www.sciencedirect.com/science/article/pii/S1871141314003011

[78] Lynch H, Leonard FC, Walia K, Lawlor PG, Duffy G, Fanning S, et al. Investigation of in-feed organic acids as a low cost strategy to combat Salmonella in grower pigs. Preventive Veterinary Medicine. 2017;**139**:50-57. Available from: https://www.sciencedirect.com/science/article/pii/S0167587716305827

[79] Zhang Y, Wang Y, Chen D, Yu B, Zheng P, Mao X, et al. Dietary chlorogenic acid supplementation affects gut morphology, antioxidant capacity and intestinal selected bacterial populations in weaned piglets. Food & Function. 2018;**9**(9):4968-4978. DOI: 10.1039/C8FO01126E

[80] Grilli E, Messina MR, Tedeschi M, Piva A. Feeding a microencapsulated blend of organic acids and nature identical compounds to weaning pigs improved growth performance and intestinal metabolism. Livestock Science. 2010;**133**(1):173-175. Available from: https://www.sciencedirect.com/science/article/pii/S1871141310002684

[81] Gao CQ, Shi HQ, Xie WY, Zhao LH, Zhang JY, Ji C, et al. Dietary supplementation with acidifiers improves the growth performance, meat quality and intestinal health of broiler chickens. Animal Nutrition. 2021;**7**(3):762-769. DOI: 10.1016/j.aninu.2021.01.005

[82] Fathi R, Samadi MS, Tabarestan R, Qotbi A, Luis A, Marín M. Effects of feed supplementation with increasing levels of organic acids on growth performance, carcass traits, gut microbiota and pH, plasma metabolites, and immune response of broilers. Animal Science Papers and Reports. 2016;**34**(2):195-206. Available from: https://www.researchgate.net/publication/301655121

[83] Pope JT, Walker GK, Rubio AA, Brake J, Jendza JA, Fahrenholz AC. Effects of corn particle size distributions and formic acid on productive and processing performance of broilers. Journal of Applied Poultry Research. 2022;**31**(4):100288. DOI: 10.1016/j.japr.2022.100288

[84] Khosravinia H, Nourmohammadi R, Afzali N. Productive performance, gut morphometry, and nutrient digestibility of broiler chicken in response to low and high dietary levels of citric acid. Journal of Applied Poultry Research. 2015;**24**(4):470-480. Available from: https://www.sciencedirect.com/science/article/pii/S1056617119303216

[85] Levy AW, Kessler JW, Fuller L, Williams S, Mathis GF, Lumpkins B,

et al. Effect of feeding an encapsulated source of butyric acid (ButiPEARL) on the performance of male Cobb broilers reared to 42 d of age. Poultry Science. 2015;**94**(8):1864-1870. DOI: 10.3382/ps/pev130

[86] Nguyen DH, Lee KY, Mohammadigheisar M, Kim IH. Evaluation of the blend of organic acids and medium-chain fatty acids in matrix coating as antibiotic growth promoter alternative on growth performance, nutrient digestibility, blood profiles, excreta microflora, and carcass quality in broilers. Poultry Science. 2018;**97**(12):4351-4358. Available from: https://www.sciencedirect.com/science/article/pii/S0032579119302652

[87] Emami K, Daneshmand A, Zafari Naeini S, Graystone EN, Broom LJ. Effects of commercial organic acid blends on male broilers challenged with E. coli K88: Performance, microbiology, intestinal morphology, and immune response. Poultry Science. 2017;**96**(9):3254-3263. DOI: 10.3382/ps/pex106

[88] Jose AM, Martin RF, Carlos LC, Ernesto AG, Jose HC, Artuo CC. Empleo de ácidos orgánicos en el agua de bebida y su efecto en el desempeño productivo en pollos de engorda. Abanico Veterinario. 2020;**10**:10-17. DOI: 10.21929/abavet2020.36

[89] Hu Z, Guo Y. Effects of dietary sodium butyrate supplementation on the intestinal morphological structure, absorptive function and gut flora in chickens. Animal Feed Science and Technology. 2007;**132**(3):240-249. Available from: https://www.sciencedirect.com/science/article/pii/S0377840106001519

[90] Lan RX, Li SQ, Zhao Z, An LL. Sodium butyrate as an effective feed additive to improve growth performance and gastrointestinal development in broilers. Veterinary Medicine and Science. 2020;**6**(3):491-499. DOI: 10.1002/vms3.250

[91] Leeson S, Namkung H, Antongiovanni M, Lee EH. Effect of butyric acid on the performance and carcass yield of broiler chickens. Poultry Science. 2005;**84**(9):1418-1422. Available from: https://www.sciencedirect.com/science/article/pii/S003257911944683X

[92] Antongiovanni M, Buccioni A, Petacchi F, Leeson S, Minieri S, Martini A, et al. Butyric acid glycerides in the diet of broiler chickens: Effects on gut histology and carcass composition. Italian Journal of Animal Science. 2007;**6**(1):19-25. DOI: 10.4081/ijas.2007.19

[93] Adil S, Banday T, Ahmad Bhat G, Salahuddin M, Raquib M, Shanaz S. Response of broiler chicken to dietary supplementation of organic acids. Journal of Central European Agriculture. 2011;**12**(3):498-508. DOI: 10.5513/Jcea01/12.3.947

[94] Youssef IMI, Mostafa AS, Ma AW. Effects of dietary inclusion of probiotics and organic acids on performance, intestinal microbiology, serum biochemistry and carcass traits of broiler chickens. Journal of World's Poultry Research. 2017;**7**(2):57-71. Available from: http://eprints.science-line.com/id/eprint/343

[95] Paul SK, Halder G, Mondal MK, Samanta G. Effect of organic acid salt on the performance and gut health of broiler chicken. The Journal of Poultry Science. 2007;**44**:389-395. DOI: 10.2141/jpsa.44.389

[96] Mikkelsen LL, Vidanarachchi JK, Olnood CG, Bao YM, Selle PH,

Choct M. Effect of potassium diformate on growth performance and gut microbiota in broiler chickens challenged with necrotic enteritis. British Poultry Science. 2009;**50**(1):66-75. DOI: 10.1080/00071660802613252

[97] Ragaa NM, Korany RMS. Studying the effect of formic acid and potassium diformate on performance, immunity and gut health of broiler chickens. Animal Nutrition. 2016;**2**(4):296-302. Available from: https://www.sciencedirect.com/science/article/pii/S2405654516301147

[98] Serrano-Gamboa MY, Arce-Menocal J, Ávila-González E, López-Coello C, Garibay-Torres L, Herrera-Camacho J. Organic acids for broilers: Effects on intestinal morphology and growth performance. Revista Colombiana de Ciencias Pecuarias. 2022;**36**(2):55-65. DOI: 10.17533/udea.rccp.v36n2a1

[99] Yesilbag D, Çolpan I. Effects of organic acid supplemented diets on growth performance, egg production and quality and on serum parameters in laying hens. Revue de Medecine Veterinaire (Toulouse). 2006;**157**:280-284

[100] Grashorn M, Gruzauskas R, Dauksiene A, Racevičiūtė-Stupelienė A, Jarule V, Miezeliene A, et al. Influence of dietary organic acids on quality and sensory attributes of chicken eggs. Archiv fur Geflugelkunde. 2013;**77**:29-34

[101] Gong H, Yang Z, Celi P, Yan L, Ding X, Bai S, et al. Effect of benzoic acid on production performance, egg quality, intestinal morphology, and cecal microbial community of laying hens. Poultry Science. 2021;**100**(1):196-205

[102] Youssef A, Hassan HMA, Ali H, Mohamed MA. Effect of probiotics, prebiotics and organic acids on layer performance and egg quality. Asian Journal of Poultry Science. 2013;**7**:65-74

[103] García J, Mandalawi HA, Fondevila G, Mateos GG. Influence of beak trimming and inclusion of sodium butyrate in the diet on growth performance and digestive tract traits of brown-egg pullets differing in initial body weight. Poultry Science. 2019;**98**(9):3937-3949. DOI: 10.3382/ps/pez129

[104] Kadim I, Al-Marzooqi W, Mahgoub O, Al-Jabri A, Al-Waheebi S. Effect of acetic acid supplementation on egg quality characteristics of commercial laying hens during hot season. International Journal of Poultry Science. 2008;**7**(10):1015-1021

[105] Abbas G, Khan SH, Rehman H. Effects of formic acid administration in the drinking water on production performance, egg quality and immune system in layers during hot season. Avian Biology Research. 2013;**6**(3):227-232. DOI: 10.3184/175815513X13740707043279

Chapter 7

The Effect of Liquid Fermented Potato Hash Diet on Testicular Size, Weight and Epididymal Semen Quality of Large White × Landrace Boars

Thivhilaheli Richard Netshirovha,
Zwivhuya Constance Raphalalani, Masindi Lottus Mphaphathi,
Mammikele Tsatsimpe, Lerato Deirdre Sehlabela
and Pulane Sebothoma

Abstract

The study aimed to evaluate testicular size, weight and epididymal semen quality of Large White × Landrace (LW × LR) boars fed fermented liquid potato hash. Diets containing either 200 (LFLPH) g/kg potato hash; 400 (HFLPH) g/kg potato hash or no fermented potato hash (control). Forty-two crossbred boars (LW × LR) weighing (25 ± 2.3 kg) were individually housed and fed ad-lib one of the seven dietary treatments for three months. Pigs were allocated to diets in complete randomized design. After 3 months and 600 ± 4 kg average body weight boars were slaughtered and epididymal semen was collected from head of epididymis. Boars that were fed control had higher ($P < 0.05$) live spermatozoa concentration than LLFPH and HFLPH diets. However, HLFPH had higher ($P < 0.05$) proximal droplet, distal droplet and dead spermatozoa concentration than control and LLFLPH. In addition, boars that were fed control and LLFPH had higher ($P < 0.05$) total, progressive, rapid motility, and lower non-progressive, static, medium and slow motility than HLFPH. It is concluded that low liquid fermented potato hash diet could be used as an alternative feed source for pigs. The results indicated that diets contain LFLPH can be used in boar diets without any adverse effects on spermatozoa quality.

Keywords: boar, epididymal semen, fermentation liquid feed, potato by-products, potato hash

1. Introduction

In animals, nutrition plays a major role in boar reproduction, including the attainment of sexual maturity, both in terms of spermatogenesis and libido.

Therefore, corn, soybean meal and wheat bran are the most common ingredients used in swine diets in South Africa. They are used because of their high nutritional value and economic benefits [1]. Pig nutrition plays an important role in the regulation of the production and reproductive efficiency of boars. In South Africa, affordability of conventional feeds has gone beyond the reach of smallholder pig farmers, due to declining grain production, increasing competition with humans for feed ingredients. To sustain local production, there is a need to look for local and readily available alternatives to substitute corn, soybean meal, fish meal, and soya oil cake that is the major protein source in pig feed. This problem could be solved by introducing less expensive local feed ingredients like potato hash in pig feed formulations to reduce the overall cost of production. Availability of agro-industrial by products has enabled smallholder pig farmers to use alternative energy sources to replace cereals in pig diets [2, 3]. Potato hash (by-product produced from the processing of chips) is produced in South Africa and can be use as alternative energy sources in pig feed. Potato by-products represent an opportunity for livestock feeders because they are an inexpensive but energy-dense dietary ingredient [4]. However, less research has been devoted to study the feeding opportunities of potato hash diets on conducive reproduction of boars. A good balance between energy and protein feeds in the ration is the key factor in achieving optimum performance. It is important to evaluate the potential of liquid fermented potato hash diet on epididymal spermatozoa quality of LW × LR pigs. Feeding fermentable carbohydrates such as raw potato starch (RPS) has been shown to strongly reduce skatole concentration in the adipose tissue of barrows [5] and entire male pigs [6, 7]. As skatole breakdown in the adipose tissue occurs rapidly, when microbial formation is reduced, feeding resistant starch during the week prior to slaughter seems enough to achieve significant reductions [8]. According to reference [9], back fat thickness is thus conceivable that selection based primarily on productive characteristics, especially for lean growth, leads to reproductive problems, such as low spermatozoa production.

Spermatozoa are produced in the testis as the result of a complex assembly line that makes a highly shaped cell, morphologically and biochemically specialized. The epididymal secretome and proteome of several mammalian species include pigs. According to reference [10] one of the changes produced in the spermatozoa through its epididymal maturation was the migration of the cytoplasmic droplet from the proximal position to the distal position of the midpiece [11]. As a result, epididymal spermatozoa and often ejaculated spermatozoa contain a heterogeneous group of spermatic cells that vary in degree of maturation and show different morphologies and fertility potential. Feeding diets containing fermented liquid feed have been shown to increase pig performance and improve the microbial environment in the gastrointestinal tract [12–14], however no work was done on epididymal semen quality of LW × LR boar. Therefore, effects of liquid fermented potato hash diet on spermatozoa traits have not been examined in LW × LR boars. To the knowledge of the authors, there is no study available, to date, on the effect liquid fermented potato hash diet on spermatozoa quality of LW × LR pigs. Several factors affect spermatozoa quality in boars; photoperiod, environmental relative humidity and temperature, nutrition, handling, breed, age, viral or bacterial infections, and, especially, the frequency of semen collection significantly affect the number of spermatozoa per ejaculate as well as spermatozoa motility and morphology [15]. To date, no study has reported the effect of liquid.

2. Materials and methods

2.1 Location and experimental area

The study was conducted at Germplasm, Conservation and Reproductive Biotechnologies Unit of Agricultural Research Council (ARC), Animal Production (AP), (ARC-AP: Irene, Pretoria, South Africa). The ARC-AP campus is located at 25°55′ South; 28°12′ East. The campus is located in the Highveld region of South Africa and situated at an altitude of 1525 m above sea level. Potato hash (PH) was collected from Simba (Isando, Gauteng, South Africa), a potato chips factory in South Africa for processing and production of fermented liquid potato hash diet (FLPH).

2.2 Fermentation process and diets

A back-slopping fermentation approach was used to prepare fermented liquid potato hash diets as described by reference [16]. Fermented liquid diets were prepared by mixing potato hash diets with water, at a ratio of 1:2. The diets were formulated to provide 14 MJ/kg digestible energy (DE), 180 g crude protein (CP)/kg and 11.6 g lysine /kg which meet and exceed the requirements of growing pigs [17]. Three diets were formulated to be isoenergetic and isonitrogenous containing either 200 (LFLPH) g/kg potato hash and 400 (HFLPH) g/kg potato hash or no fermented potato hash (control). The seven dietary treatments are showed in **Figure 1**: **A**-CON (control diet non-fermented, contain no potato hash), **B**-LFC (liquid fermented control diet), **C**-LLPH (diet containing 200 g potato hash.kg^{-1} as fed), **D**-HLFPH (diet containing 400 g potato hash.kg^{-1} as fed), **E**-LFCE (fermented control diet treated with an exogenous xylanase enzyme (Natugrain TS L®), **F**-LLFPHE (diet containing 200 g potato hash.kg^{-1} as fed treated with an exogenous xylanase enzyme (Natugrain TS L®)), **G**-HLFPHE (diet containing 400 g potato hash.kg^{-1} treated with an exogenous xylanase enzyme (Natugrain TS L®). The fermented diets were stored in closed 100 L drums under agitation at 25°C for 8 hours before being fed to the pigs. Pigs were adapted to diets for a period of ten days. Both experimental diets and water were provided *ad* libitum for two months (**Table 1**) [16].

2.3 Characterization of the liquid fermented potato hash

Fermented liquid feed is defined as a mixture of feed and water that is stored in a tank at a specific temperature and for a specific time before being fed to animals [16]. Fermented liquid by-products, fermented diets can also be achieved when dry

Figure 1.
Dietary treatments of liquid fermented potato hash; A- CON- control (liquid fermented control with and without enzyme); LLFPH - low inclusion of liquid fermented potato hash with and without enzyme; HLFPH - high inclusion of liquid fermented potato hash with and without enzyme; LFC + E-liquid fermented control with enzyme diet; LFC – Liquid fermented control diet.

	Control	LLFPH	HLFPH
Experimental diets[1]			
Ingredient kg			
Hominy Chop	608.7	504.4	400
Molasses	20	15	10
Potato Hash	0	200	400
Soybean Oilcake	181.4	166.7	152.1
Maize meal	150	75	0
Monocalcium Phosphate	5	8.1	11.2
Limestone	18.8	16.3	13.7
Lysine HCI	8	6.5	5
Salt	4	4	4
Vitamin-mineral Premix[2]	4	4	4
Calculated composition			
Nutrients g/kg			
DM	892	605	599
Ash	2.5	31	37
CP	180	180	180
Crude fiber	57	58	60
Calcium	9.12	9.12	9.12
Phosphorus	5.47	5.46	5.46
Lysine	11.6	11.6	11.6
Methionine	0.67	0.56	0.56
DE MJ/kg	13.5	13.5	13.5

[1]*CON = control (liquid fermented control with and without enzyme); LLFPH = low inclusion of liquid fermented potato hash with and without enzyme; HLFPH = high inclusion of liquid fermented potato hash with and without enzyme.*
[2]*Provided the following per kg of diet: 6500 IU vitamin A, 1200 IU vitamin D3, 40 IU vitamin E, 2 mg vitamin K3, 1–5 mg vitamin B1, 4.5 mg vitamin B2, 0.03 mg vitamin B12, 2.5 mg vitamin B6, 25 mg niacin, 12 mg calcium pantothenate, 190.5 mg choline, 0.6 mg folic acid, 0.05 mg biotin, 40 mg manganese, 100 mg zinc, 125 mg copper, 1 mg iodine, 100 mg ferrous, and 0.3 mg selenium [16].*

Table 1.
Composition and chemical analysis of the diet on as-is basis of different inclusion levels of liquid fermented potato hash.

compound feed is mixed with water and stored for at least 8 hours [18]. This processing method can easily be used under small farmer conditions. It is also known as soaking. A short-term perseveration technique is used to store liquid by-products. Feeding fermented liquid compound feeds to weaned piglets improved daily gain and changed the gastrointestinal environment in a more desirable direction compared to non-fermented liquid feeds [14]. Feeding fermented liquid compound feeds to weaned piglets improved daily gain and changed the gastrointestinal environment in a more desirable direction compared to non-fermented liquid feeds. Processed potatoes such as hash represent a potential energy source that could replace or be included in the traditional pig diet [13]. Another disadvantage of feeding hash to growing pigs is that it contains a high-fiber diet that increases the passage rate in growing pigs [16].

2.4 Pigs, experimental design and housing

Forty-two (6 pigs per treatment) crossbred boars (Large White × Landrace) aged 55 days with an average weight of 25.5 ± 3 kg was randomly selected from the ARC-AP Irene, pig breeding unit. The boars were allocated to dietary treatments in a completely randomized design. The boars were housed individually in 1.54 × 0.8 m pens in environmentally controlled houses with the temperature ranging from 22 to 25°C. Each pen was provided with wood shaving. Daily feed offered and weekly orts were recorded. Orts were dried, weighed and discarded daily. Weights of feed refusals and orts were subtracted from the total weight of the feed allocated to determine feed intake for that week. Weight of the feed consumed each week was divided by seven to determine the average daily feed intake. Feed was supplied ad libitum and water was made available at all times through drinking nipples. Mortality and morbidities were noted. Morbidities were diagnosed and the necessary treatments were done [16].

2.5 Boars slaughtering and measurements of testicular traits

These pigs were humanely slaughtered when they attained a weight of 60 ± 4 kg. The pigs were then stunned with an electrical stunner set at 220 V and 1.8 A with a current flow for 6 s and exsanguinated within 10 s of stunning [9]. While boar was bleeding, testis and epididymis were carefully removed before the carcass was dipped into hot water, de-haired and eviscerated. The length, width and weight for left and right testis were measured. The length, width and weight for left and right testis were measured using caliper. The weight of testis was measured using a sensitive digital weighing scale. The right and left epididymis was trimmed off the body of the testis. Scrotal sacs were incised to exteriorize the testis and epididymides carefully collected, trimming off adhering tissues and weighed using a sensitive electronic balance. Other testicular and epididymal morphometric characteristics such as length, diameter, and volume of the testis, and epididymal length were also measured. The testis length, width and epididymal length were measured with the aid of a pair of Vernier calipers, while the testis volume was measured by water displacement according to Archimedes principle. Paired and mean testicular and epididymal parameters were computed from data for left and right testis and epididymis.

2.6 Semen sample collection and evaluation

The boars were slaughtered, and semen samples were collected from head of the epididymis through a razor blade incision made by on the right and left testicles. The epididymal semen samples were collected into a graduated 15 mL tube. The semen samples were placed in well-insulated flasks maintained at a warm temperature (37°C) before being transported to the laboratory within 30 minutes for measurement of semen traits categorized into microscopic evaluation.

2.7 Spermatozoa motility

Spermatozoa motility was determined using a sperm class analyzer® (CASA) (Microptic S.L, Spain). Five hundred microliters of Ham's F-10 (Sigma-Aldrich, South Africa) and 5 μL of semen were mixed in a 1 mL graduated tube and incubated for 5 minutes at 37°C. After incubation, 10 μL of extended semen was placed on a pre-warmed microscopic slide (37°C), mounted with a cover slip and examined (×10) under a phase contrast microscope (Nikon, Japan). Spermatozoa motility was categorized as follows: Progression (%), Total motility (TM) - is a sum of progressive and non-progressive motility; Progressive motility (PM) - spermatozoa that are moving forward; Average values of velocity parameters; Curvilinear velocity (VCL) - average velocity which measures a spermatozoa movement along its actual path (μm/s); Straight-line velocity (VSL) - average velocity which measures a spermatozoa movement along a straight line from beginning to the end (μm/s); Average path velocity (VAP) - average velocity of the smoothed cell path (μm/s); Linearity (LIN) - linearity movement is a ratio of VSL/VCL (%); Straightness (STR) - straight line movement is a ratio of VSL/VAP (%) and Wobble (WOB) - wavering movement which is a ratio VAP/VCL (%) [19–21].

2.8 Spermatozoa concentration

Spermatozoa concentration was determined with a 6310 spectrophotometer (Jenway, United Kingdom). A square cuvette was filled with 3 mL of sodium citrate solution and placed in a spectrophotometer for at least 30 seconds. Raw semen (15 μL) was added in a square cuvette containing the sodium citrate solution, again placed in a spectrophotometer in order to read the absorbance. The absorbance read was used to determine the final spermatozoa concentration (10^6/mL) with the aid of a formula (201 × 25.97 × absorbance - 0.3). The final spermatozoa concentration was recorded in millions per milliliter spermatozoa concentration. Semen pH was measured using the pH meter (Oaklon, EW35614–30, ColeParmer, East Bunker Court, Vernon Hills, IL, USA) [20].

2.9 Spermatozoa morphology

The morphology was determined microscopically after staining the semen samples with Eosin Nigrosin stain (Onderstepoort, Pretoria) on a slide. Boar semen was added to 20 μL Eosin Nigrosin staining solution in a 0.6 mL micro-centrifuge graduated tube and mixed gently. A drop of 5 μL boar semen and Eosin Nigrosin stain was placed on a clear end of a microscope slide and smeared. Semen samples was determined using Eosin Nigrosin stain (pH - 8.39), to determine percentage live or dead spermatozoa and evaluation of the spermatozoa morphology (normal or abnormal). The spermatozoa smears were prepared on a clean, warm microscope slide to avoid temperature shock to the spermatozoa and evaluated on the same day of semen collection and with the aid of a fluorescent microscope (BX 51TF) using an oil immersion objective (×100 magnification). Live spermatozoa were further evaluated for spermatozoa morphology and abnormalities. Abnormalities of the spermatozoa were categorized as primary (small, large or swollen head, double heads, abnormal acrosome, elongated and mid-piece, double and short tail), secondary (detached, loose or damaged acrosomes, bent and protoplasmic droplets of the mid-piece, bent and shoe-hook tail) and tertiary abnormalities (reacted acrosomes and coiled tails), such as live, dead, distal droplet, head, midpiece and tail [21].

2.10 Membrane integrity

Membrane integrity (**Figure 2**) was assessed using the osmotic resistance test (the hypoosmotic swelling test – HOST) by incubating an aliquot (100 μL) of semen

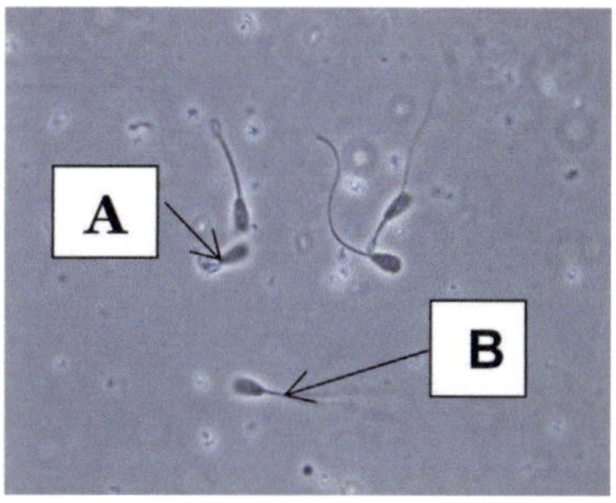

Figure 2.
Membrane integrity of the raw boar spermatozoa evaluated with HOST. (A) Spermatozoa with intact membrane and (B) Spermatozoa with damaged membrane.

sample with 1 ml of double distilled water at 37°C for 30 minutes [22]. After incubation, a pinch of Eosin was added; a drop of the well-mixed sample was placed on a glass slide and covered with cover slip. This slide was observed at 400× magnification under the phase contrast microscope. Spermatozoa with swollen tail were counted as HOST positive. A minimum of 200 spermatozoa were observed for tail coiling (**Figure 2**). The percentage of reactive spermatozoa was then calculated by subtracting the percentage of tail defects recorded in the sperm population before incubation in HOST media was carried out.

2.11 Statistical analysis

The liquid fermented potato hash and genotype on epididymis spermatozoa quality, testicular development was performed using (SAS) version 9.3 statistical software (SAS, 1999). The GLM procedure was also used to determine the effect of LLFPH, HLFPH, LFC + E, LFC, LLFPH+E, HLFPH+E and genotype. A 5% significance level was used.

3. Results

3.1 Epididymal semen volume, semen pH, spermatozoa concentration, and abnormalities spermatozoa morphology

The effect of supplementation liquid fermented potato hash diet on epididymal semen volume, spermatozoa concentration, semen pH and abnormalities spermatozoa morphology, in LW × LR boars are shown in **Figure 3**. However, LFC and HLFPH+E had lower ($P < 0.05$) epididymal semen volume compared to control, LLFPH, HLFPH, LFC + E and LLFPH+E treatments. There was no difference ($P > 0.05$) in epididymal semen pH between the treatments. There was a difference ($P < 0.05$) in epididymal spermatozoa concentration between the treatments. Boars fed HLFPH+E had lower spermatozoa concentration compared to boars consuming control, LLFPH, HLFPH, LLFPH+E, LFC + E and LFC treatments. There was a difference ($P < 0.05$) in epididymal head abnormalities spermatozoa between the treatments. Where boars fed LLFPH had higher head abnormalities spermatozoa compared to control, HLFPH, LLFPH+E, HLFPH+E, LFC + E and LFC treatments. There was a difference ($P > 0.05$) in epididymal tail abnormalities spermatozoa

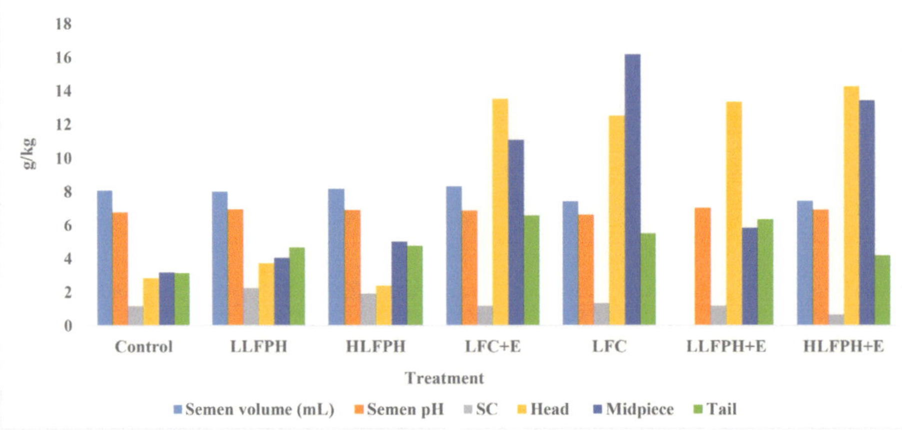

Figure 3.
The effect of supplementation liquid fermented potato hash diet on epididymal semen volume, spermatozoa concentration, semen pH and abnormalities spermatozoa morphology, in LW × LR boar (±SEM). [abc]Values with different superscripts within a row differ significantly (P < 0.05, SEM- standard error of mean, SC – Sperm concentration; CON- control (liquid fermented control with and without enzyme); LLFPH - low inclusion of liquid fermented potato hash with and without enzyme; HLFPH - high inclusion of liquid fermented potato hash with and without enzyme; LFC + E-liquid fermented control with enzyme diet; LFC - Liquid fermented control diet.

between the different treatments. Boars fed control treatment had lower epididymal tail abnormalities spermatozoa compared to boars on the other dietary treatments.

3.2 The effect of supplementation liquid fermented potato hash diet on epididymal spermatozoa parameters

The mean supplementation liquid fermented potato hash diet on epididymal spermatozoa motility and velocity parameters of LW × LR boar semen as measured by CASA are shown in **Table 2**. There was a difference (P < 0.05) in total and progressive spermatozoa motility between the treatments. Boars fed LLFPH had higher epididymal total motility spermatozoa and progressive motility compared to boars on all the other dietary treatments. Boars fed LFC had epididymal lower progressive motility compared to the control, LLFPH, HLFPH, HLFPH+E, LLFPH+E and LFC + E treatments. There was a difference (P < 0.05) in both rapid and slow epididymal spermatozoa between the treatments., Boars fed HLFPH+E had lower epididymal rapid spermatozoa compared to control, LLFPH, HLFPH, LLFPH+E, LFC + E and LFC treatments. There was an increase in slow spermatozoa with the increases of HLFPH+E treatments. There was no difference (P > 0.05) in VCL spermatozoa between the treatments. However, pigs fed HLFPH+E tended to have decreased VCL spermatozoa. There was a difference (P < 0.05) in epididymal VSL spermatozoa between the treatments. However, pigs fed control (35.08) had higher values VSL epididymal spermatozoa compared to LLFPH, HLFPH, HLFPH+E, LLFPH+E, LFC + E and LFC (27.93, 24.53, 26.55, 32.15, 25.70 & 23.62) treatments. There was no difference (P > 0.05) in epididymal VAP spermatozoa between control, LLFPH, HLFPH, LLFPH+E, LFC + E and LFC but there was difference (P < 0.05) with HLFPH+E treatments. There was no difference (P > 0.05) in epididymal linearity between the

Parameters	Control (n = 6)	LLFPH (n = 6)	HLFPH (n = 6)	LFC + E (n = 6)	LFC (n = 6)	LLFPH + E (n = 6)	HLFPH + E (n = 6)	SEM	P-value
TM (%)	80.68^{a}	83.68^{a}	78.47^{a}	69.02^{bc}	64.73^{c}	74.03^{b}	78.58^{a}	1.989	<.0001
PM (%)	64.40^{b}	70.93^{a}	69.62^{a}	57.45^{bc}	40.07^{c}	57.02^{bc}	63.67^{b}	3.391	<.0001
RAP (%)	72.37^{a}	58.53^{ab}	47.62^{bc}	52.15^{b}	57.45^{b}	60.70^{ab}	49.43^{c}	5.055	0.0271
SLW (%)	2.67^{c}	3.48^{c}	4.68^{bc}	5.52^{ab}	4.72^{bc}	3.10^{c}	7.03^{a}	0.639	0.0003
VCL (μm/s)	138.00^{a}	131.75^{a}	140.55^{a}	134.24^{a}	113.93^{a}	129.09^{a}	114.63^{a}	11.470	0.5301
VSL (μm/s)	35.08^{a}	27.93^{ab}	24.53^{b}	26.55^{bc}	32.15^{a}	25.70^{bc}	23.62^{c}	2.308	0.0108
VAP (μm/s)	65.98^{a}	57.80^{a}	56.47^{a}	57.12^{a}	58.75^{a}	59.38^{a}	45.92^{b}	4.401	0.1221
LIN (%)	25.78^{a}	22.13^{a}	22.75^{a}	19.75^{a}	25.25^{a}	20.08^{a}	26.92^{a}	2.808	0.4367
WOB (%)	46.30^{a}	37.10^{b}	42.77^{a}	42.43^{a}	46.30^{a}	43.15^{a}	43.17^{a}	2.660	0.2646
ALH (%)	4.47^{a}	4.47^{a}	3.93^{a}	3.90^{a}	4.40^{a}	3.77^{a}	4.27^{a}	0.275	0.2993
BCF Hz	8.40^{b}	7.17^{b}	9.08^{b}	12.67^{a}	11.00^{a}	8.95^{b}	8.17^{b}	1.109	0.0229
MED (%)	20.45^{b}	19.68^{b}	31.00^{a}	15.70^{bc}	11.60^{bc}	12.37^{bc}	5.72^{c}	2.289	<.0001
STR (%)	49.08^{a}	49.58^{a}	46.88^{b}	46.67^{b}	54.75^{a}	45.22^{b}	53.58^{a}	1.969	0.0106
Static (%)	5.167^{c}	18.15^{bc}	24.30^{b}	26.63^{b}	26.23^{b}	23.83^{b}	37.82^{a}	4.459	0.0010
Live spermatozoa (%)	81.03^{a}	80.68^{a}	80.80^{a}	52.83^{c}	59.83^{bc}	70.67^{b}	64.50^{b}	1.679	<.0001
Dead spermatozoa (%)	10.02^{b}	7.78^{b}	7.37^{b}	15.17^{a}	6.00^{b}	3.83^{c}	3.67^{c}	0.915	<.0001
HOST (%)	76.98^{a}	76.43^{a}	80.00^{a}	75.42^{a}	80.30^{a}	64.50^{b}	65.65^{b}	2.911	0.0009

^{abc}*Values with different superscripts within a row differ significantly ($P < 0.05$). TM - Total motility, PM - Progressive motility, SLW - Slow, MED - Medium. RAP - Rapid, VCL - Velocity curvilinear, VSL - Velocity straight line, VAP - Velocity average pathway, LIN - Linearity, STR - Straightness, WOB - Wobble, ALH - Amplitude, BCF- Frequency, CON - control (liquid fermented control with and without enzyme), LLFPH - low inclusion of liquid fermented potato hash with and without enzyme, HLFPH - high inclusion of liquid fermented potato hash with and without enzyme and HOST - hypo-osmotic swelling test.*

Table 2.
Influence of liquid fermented potato hash inclusion on semen and spermatozoa characteristics of LW × LR boars (+SEM).

treatments. However, pigs fed LLFPH+E had lower epididymal linearity spermatozoa compared to control, LLFPH, HLFPH, HLFPH+E, LFC + E and LFC treatments. There was no difference (P > 0.05) in epididymal wobble spermatozoa between the control, HLFPH, HLFPH+E, LLFPH+E, LFC + E and LFC treatments, however there were difference with LLFPH treatment. There was no difference (P > 0.05) in ALH epididymal spermatozoa between the treatments. There was no difference (P > 0.05) in BCF spermatozoa between the treatments. However, pigs fed LFC + E had higher BCF spermatozoa compared to control, LLFPH, HLFPH, LLFPH+E, HLFPH+E, and LFC treatments. Pigs fed control diet had higher medium spermatozoa compared to the LFC, LLFPH, HLFPH, HLFPH+E, LLFPH+E and LFC + E treatments. There was a difference (P < 0.05) in straightness spermatozoa between the treatments. However, pigs fed LFC had higher epididymal straightness spermatozoa compared to the control, LLFPH, HLFPH, HLFPH+E, LLFPH+E and LFC + E treatments. There was a difference (P < 0.05) in statics spermatozoa between the treatments. However, pigs fed HLFPH+E had an increase in epididymal statics spermatozoa compared to the control, LFC, LLFPH, HLFPH, LLFPH+E and LFC + E treatments.

3.3 The effect of supplementation liquid fermented potato hash diet on epididymal morphology and membrane integrity parameters

Analysis of boar spermatozoa morphology with Eosin/Nigrosin staining solution viewed under a fluorescence microscope at 100x magnification is indicated in **Figure 4**. Boars fed LFC + E had lower live spermatozoa compared to control, LLFPH, HLFPH, LLFPH+E, HLFPH+E and LFC treatments. There was a difference (P < 0.05) in dead epididymal spermatozoa between the treatments. Although pigs fed LFC + E had higher dead epididymal spermatozoa compared to the control, LFC, LLFPH, LLFPH+E, HLFPH and LLFPH+E treatments. Membrane integrity of the raw boar spermatozoa evaluated with HOST shown in **Figure 3**. The HLFPH and LFC had a higher (P < 0.05) osmotic swelling test compared to control, LLFPH, LLFPH+E, LFC + E and LLFPH+E treatments.

3.4 Testicular and epididymal morphometric

Testicular and epididymal morphometric characteristics of White × Landrace boars fed liquid fermented potato hash diets are show in **Table 3**. There was a difference (P < 0.05) in the right testis weight and width indexes between the treatments.

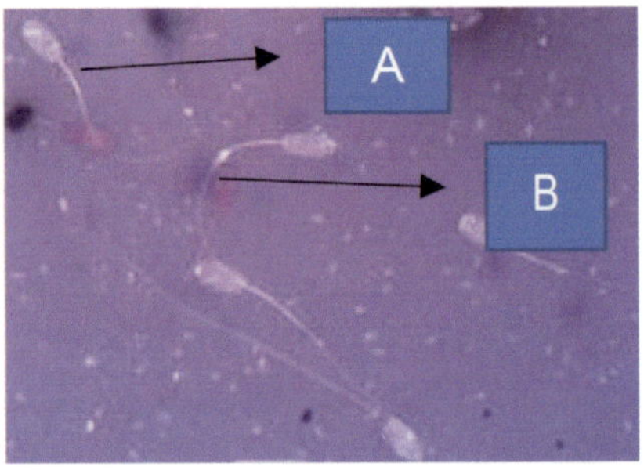

Figure 4.
Analysis of boar spermatozoa morphology with eosin/Nigrosin staining solution viewed under a fluorescence microscope at 100x magnification. (A): Live spermatozoa and (B): Dead spermatozoa.

DOI: http://dx.doi.org/10.5772/intechopen.113120

Variables	Control (n = 6)	LLFPH (n = 6)	HLFPH (n = 6)	LFC + E (n = 6)	LFC (n = 6)	LLFPH + E (n = 6)	HLFPH + E (n = 6)	SEM	P-value
Right testis weight index (kg)	0.96^{a}	0.86^{ab}	0.74^{b}	0.99^{a}	0.88^{ab}	0.88^{ab}	1.00^{a}	0.032	<.0001
Right testis width index mm	2.07^{a}	1.81^{ab}	1.50^{b}	2.22^{a}	2.00^{a}	1.81^{ab}	2.20^{a}	0.098	<.0001
Left testis weight index (kg)	0.95^{a}	0.85^{ab}	0.73^{b}	0.98^{a}	0.86^{ab}	0.87^{b}	0.98^{a}	0.031	<.0001
Left testis width index (mm)	2.00^{b}	1.77^{bc}	1.46^{c}	3.03^{a}	2.00^{b}	1.78^{bc}	2.15^{b}	0.350	0.0900
Right testis length index (mm)	0.97^{a}	0.86^{bc}	0.73^{c}	0.96^{a}	1.00^{a}	0.90^{a}	0.97^{a}	0.026	<.0001
Left testis length index, (mm)	0.98^{a}	0.85^{ab}	0.74^{b}	0.95^{a}	0.98^{a}	0.88^{ab}	0.96^{a}	0.025	<.0001
Right epididymis length (mm)	0.53^{a}	0.47^{a}	0.37^{b}	0.53^{a}	0.50^{a}	0.44^{b}	0.49^{a}	0.018	<.0001
Right epididymis weight index (g/kg)	3.21^{a}	2.94^{a}	3.26^{a}	2.91^{a}	2.87^{a}	2.20^{b}	2.46^{ab}	0.268	0.0830
Left epididymis length (mm)	0.53^{a}	0.46^{ab}	0.36^{b}	0.51^{a}	0.49^{a}	0.44^{b}	0.48^{a}	0.018	<.0001
Left epididymis weight index (mm)	2.91^{a}	2.75^{a}	3.04^{a}	2.88^{a}	2.87^{a}	2.21^{b}	2.46^{ab}	0.241	0.2138

[ab]*Means on same row with different superscripts differ significantly ($P < 0.05$), SEM - standard error of mean, C - control (liquid fermented control with and without enzyme); LLFPH - low inclusion of liquid fermented potato hash with and without enzyme; HLFPH - high inclusion of liquid fermented potato hash with and without enzyme; LFC + E - Liquid fermented control with enzyme diet; LFC - Liquid fermented control diet.*

Table 3.
Influence of dietary liquid fermented potato hash inclusion on testicular development for white × landrace boars (+SEM).

Boars fed HLFPH+E had higher right testis weight index compared to the control, LLFPH, HLFPH, LFC + E, LLFPH+E and HLFPH+E treatments. Pigs fed HLFPH had lower right testis width index compared to the control, LLFPH, LFC + E, LFC, LLFPH+E, and HLFPH+E treatments. The LLFPH+E and LLFPH had a higher ($P < 0.05$) left testis weight index compared to control, HLFPH, LFC + E, LFC, and HLFPH+E treatments. Boars fed LFC + E had higher left testis width index compared to the control, LLFPH, HLFPH, LFC, LLFPH+E and HLFPH+E treatments. However, there was no difference ($P > 0.05$) in left testis width and length index between the treatments. Boars fed HLFPH had the lowest right testis length index compared to all the other treatments. There was a difference ($P < 0.05$) in right epididymis length between the treatments, with boars consuming HLFPH and LLFPH diets having the shortest right epididymis length. There was a difference ($P < 0.05$) in left epididymis length between the dietary treatments, demonstrated by the shortest left epididymis in boars that were fed HLFPH and LLFPH+E diets.

4. Discussion

4.1 The effect of supplementation liquid fermented potato hash diet on semen volume, pH, spermatozoa concentration and sperm motility parameters

The study hypothesized that epididymal spermatozoa quality was affected by inclusion levels of liquid fermented potato hash supplementation in LW × LR boars. Supplementation of liquid fermented potato hash in our study did not show any significant negative effect on semen volume of LW × LR pigs. Similarly, Ogunlade et al. [1] reported that dietary supplementation with *Saccharomyces cerevisiae* to layer breeders improved semen quality by increasing semen volume, sperm concentration, and motility and by reducing dead and abnormal sperm as opposed to the untreated group. There were no significant differences in the epididymal semen volume were recorded in this study. Similarly, Santos et al. [23] who observed no effect of palm kernel cake (PKC) and coconut meal (CM) diets on semen volume, gross motility, vigor, and spermatozoa morphologic defects of the ejaculates in water buffalo semen. Wähner et al. [24] also reported an increase in semen volume and higher spermatozoa concentration, as l-carnitine increased in boars. In our study the inclusion levels of liquid fermented potato hash did not show any significant effect on spermatozoa concentration [25]. Similar results were obtained with male buffalo calves, in which the supplementation of yeast fermentation product (0%,0.5%, or 1.0%) of the diet did not impact sperm concentration, volume, motility, and viability [18]. A significant increase was recorded in the epididymal sperm concentration and motility of rats consuming fermented rooibos and 'green' rooibos when compared with the control group and other experimental groups [26]. No significant difference in the epididymal head abnormalities spermatozoa was observed in contrast to the findings of [2, 3] that no effect of (200, 400 and 600 mg/kg of sweet potato (*Ipomoea batatas*) leaf extract as compared to the control diet on spermatozoa concentration on semen in rabbit.

Amao and Showunmi [27] reported that rabbit increases epididymal spermatozoa concentration in bucks fed control and fermented cottonseed cake than bucks fed raw cottonseed cake-based diet. Similarly, Chung et al. [28] found that *Lepidium meyenii* (Maca) aqueous extract increased the epididymal sperm count of a rat. Ekpo et al. [29] reported that the recent evidence indicates that there was no change in epididymal spermatozoa parameters from the rats fed sweet potato. In the present study, semen pH was not affected by liquid fermented potato hash supplementation in Large White x Landrace pigs was expected. Ekpo et al. [29] reported that increases of epididymal semen pH, as inclusion levels of sweet potato increased in albino rats. The semen pH of the ejaculates ranged between 6.8 and 7.0 with no difference among the palm kernel cake (PKC) and coconut meal (CM), also in accordance to previous findings Sansone et al. [30, 31]. According to Uno et al. [32] there was no significant difference in the epididymal semen pH among the different treatment groups. The pH of the semen was between the range of 7.18 to 7.25 in the current study as shown in **Table 2**. Uno et al. [2, 3] reported a significant decrease in the weight of epididymal spermatozoa count in albino rats fed/supplemented with the leaves extract. Amao and Showunmi [27] found that spermatozoa count increased with fermented cottonseed cake diet. While Etchu et al. [33] found that semen volume, spermatozoa concentration and sperm output decreased with processed sweet potato in diet. An increase in total motility spermatozoa suggests that including liquid fermented potato hash beyond (diet containing 200 g potato hash.kg^{-1} diet) compromises motility of spermatozoa. An increase in total motility spermatozoa found in the current study correspond with finds reported by Amao and Showunmi [27] of the highest spermatozoa motility in rabbit supplemented with 200 mg fermented cottonseed cake. Pant et al. [34] and dos Santos et al. [23] indicated that progressive spermatozoa motility was higher in water buffalo fed with a diet containing 69.3% palm kernel cake. In addition, Rubus coreanus has been reported to increase spermatozoa counts and motility in white rabbit [10]. The increase in sperm output by testis size line boars was due to larger testis and greater rates of daily sperm production [35]. The observation that progressive motility was affected by inclusion levels of liquid fermented potato hash in LW × LR boars. Dos Santos et al. [23] reported that progressive spermatozoa motility of buffalo with supplemented palm kernel cake was increased than the rabbit fed control group. Zhao et al. [36] reported that, these results suggested that wine grape pomace could be used as a feed ingredient in rams to alleviate restraint induced oxidative stress and improve epididymal spermatozoa quality. Etchu et al. [33] reported that semen volume, spermatozoa concentration (109/cm^3), spermatozoa motility (%), live/dead ratio (%), spermatozoa output (109/cm^3), abnormal spermatozoa (%), semen pH, birds fed fermented sweet potato-based diets was decreases than the birds fed control diet, sliced potatoes, and grated potatoes. McDaniel et al. [37] reported that the effect of yeast fermentation product on the semen quality index was mathematically due to a reduction in spermatozoa motility in White Leghorn roosters. In White Leghorn roosters, [38] cited that that spermatozoa motility of White Leghorn roosters supplemented with fermentation product to roosters linearly decreased possibly due to the linear increase in the number of bacteria per spermatozoa and yeast per spermatozoa. Uno et al. [2, 3] who observed significant effect of on the weight of epididymes, spermatozoa motility and spermatozoa viability in albino rats treated the leaves extract. Similarly, van Dorland et al. [39] suggested that the supplementation of yeast product can improve semen quality in horses by increasing the antioxidant capacity in the semen. However, mammals and birds exhibit remarkable differences in their reproductive systems, so it is possible that the yeast benefits reported in rats would not apply to avian species. Abaza et al. [40] reported that dietary supplementation with *Saccharomyces cerevisiae* to layer breeders improved semen quality by increasing semen volume, sperm concentration, and motility and by reducing dead and abnormal sperm as opposed to the untreated group.

4.2 The effect of supplementation liquid fermented potato hash diet on spermatozoa abnormalities

An increase in live spermatozoa suggests that including potato hash beyond control diet (not fermented and without potato hash) compromises livability of spermatozoa. An increase in live spermatozoa found in the current study correspond well with finds by Ragab et al. [41] who reported a highest total motile, total live and total normal sperm in rabbits supplemented with 2.5 g pumpkin seed oil/kg diet plus 2.5 g black seed oil/kg diet /kg diet (PSO + BSO). Ekpo et al. [29, 42] indicated that number of abnormal and head spermatozoa was lower in albino rats fed with a diet containing 200 mg/kg of sweet potato. Findings of Bréque et al. [43] does not support the outcomes of the present study that increasing levels of liquid fermented potato hash decreased dead spermatozoa and head abnormalities. A study by Lovercamp et al. [19] regarding farrowing rate revealed that boars with a low fertilization performance had a significantly lower pro-portion of normal spermatozoa than boars with a high performance. Njoku et al. [44] highlighted that an addition of pumpkin seed reduced considerably the percentage of morphologically changed spermatozoa while the sperm count, motile sperm and viability improved. Skoracka et al. [45] reported that high levels of dietary zinc supplementation preservation sperm morphology, sperm count and function, and thus, for the proper course of fertilization. It worth noting that liquid fermented potato hash requirement for pigs, although some of these gave good results, it is difficult to obtain a reliable and consistent fermentation due to some factors.

Supplementation of liquid fermented potato hash diet in our study did not show any significant effect on live and dead spermatozoa. No significant difference in the spermatozoa live and head abnormalities was similar to the findings of Ekpo et al. [29] who observed no effect of inclusion levels of sweet potato (*Ipomoea batatas*) on spermatozoa live and head abnormalities in male albino rats. Etchu et al. [33] reported that there was a decrease of live/dead ratio (%), abnormal spermatozoa (%) of birds fed fermented sweet potato-based diets than birds fed control diet, sliced potatoes, and grated potatoes. Gofura et al. [20] reported that the abnormality of spermatozoa increased with dietary of purple sweet potato increased in rats. Morphologically normal spermatozoa (50.67%) were also significantly lower than that of the fermented cottonseed cake group but similar to that of the control [27]. However, non-motile spermatozoa, abnormal spermatozoa, round and elongated spermatids were not significantly affected by fermented cottonseed cake diets [27]. Dietary supplementation of yeast fermentation product leads to a linear decrease in the spermatozoa quality index, which is indicative of overall semen quality and is affected by spermatozoa viability, concentration, and motility [37]. The results obtained by Uno et al. [2, 3] revealed a significant decrease in the spermatozoa viability and spermatozoa count while spermatozoa head abnormalities significantly increased in albino rats treated the leaves extract.

4.3 The effect of supplementation liquid fermented potato hash diet on testicular, epididymal weight and size

An increase in testicular weight as supplementary PH levels increased is similar to findings of Ekpo et al. [29] higher testis weights and epididymes testis length and width in rats supplemented 600 mg/kg of sweet potato (*Ipomoea batatas*) than others group and control diet. In contrast, increasing levels of leaf meal inclusion resulted in decreased testicular size of rabbits [46]. Testicular weights have been reported to have a high correlation with sperm reserve in the testis or epididymis and therefore a reflection of sperm production [46]. In rabbits, Ansa et al. [47] find that testis length, testis circumference, testis weight, testis volume, epididymis weight of rabbit supplemented with 900 mg kg^{-1} methanolic extract of *Phoenix dactylifera* fruit per day was bigger than the un-supplemented group. Peerry and Petterson [48] reported that size, length, and width of testis are good indicators of present and future sperm production. In the present study, weight epididymis was affected by dietary levels of liquid fermented potato hash supplementation in LW × LR pigs. An increase on right testis weight index as supplementary levels HLFPH+E increased is similar to findings of Adienbo and Wodu [49] where sweet potato supplementation increased weights of the testis and epididymes in animals. In contrast Uno et al. [2, 3] reported that testis weight was not affected by sweet potato (*Ipomoea batatas*) 200, 400 and 600 mg/kg when compared with the control diet in rats. Amao and Showunmi [27] reported that bucks fed on raw cottonseed cake -based diet had significantly higher values for left, right and mean epididymal weights than for bucks on other fermented cottonseed cake. Epididymis stores and transports spermatozoa that are produced in the testis of boars [50].

An increase in epididymis weight was expected. Testicular testosterone output is primarily regulated by the pulsatile pattern of pituitary LH secretion [51]. These androgenic properties enhance concentration of luteinizing hormone, which is responsible for testicular development such as epididymis and testicular weight [50]. An increase in epididymis weight could be is associated with quadratic increase in semen volume. Amao and Showunmi [27] reported increased epididymides (left, right and mean) when rabbit bucks were fed fermented cottonseed cake compared to those fed control diet. In addition, Hyacinth et al. [21] reported a decrease on the morphometric, weights of testis and epididymides with lower inclusion of tephrosia bracteolate leaf meal. Bitto et al. [52] reported that a decreased on paired of testis weight of cockerels fed supplementation of cassava peel meal up to 30% in diet. Majid et al. [53] reported an increased size, weight, and relative weight of testis and epididymis of rats administered with 300 mg/kg day of sweet potato.

5. Conclusions

From the results of the present study the following main conclusions can be drawn:

1. liquid fermented low potato hash diet could be an alternative feed source for growing boars.

2. Therefore, the results indicated that diet contain LFLPH can be used in growing boars without any adverse effects on semen quality of boars.

3. It is recommended that further investigations on fed fermented liquid potato hash on semen collected from living boars should evaluated for semen quality, testicular morphology, and histology.

4. The findings showed that LFLPH-containing feeds may be utilized in pig diets without having any negative effects on the quality of spermatozoa.

Acknowledgements

The authors would like to express gratitude to the Agricultural Research Council (ARC) - Nutrition and Germplasm Conservation and Reproductive Biotechnologies (GCRB) staff for the assistance and Gauteng Department of Agriculture and Rural Development (GDARD). Dedication goes to the late Dr. R.S Thomas who assisted with providing funding from GDARD), and the staff at the Agricultural Research Council-Irene who assisted with the management of pigs throughout the trial are also acknowledged.

Conflict of interest

The authors declare no conflict of interest.

Ethical approval

The experimental procedures used in this study were in accordance with guidelines of the Agricultural Research Council Animal Ethics committee (Reference: APIEC16/037), during the period from November 2016 to August 2017.

Abbreviations

ALH	amplitude of lateral head displacement, µm
ARC	Agricultural Research Council
AP	Animal production
BCF	beat-cross frequency, Hz
BSO	black seed oil
CASA	Computer-Assisted Sperm Analysis
CON	control diet
CP	crude protein
CM	coconut meal
DE	digestible energy
SC	sperm concertation, 106/mL
GCRB	Germplasm Conservation and Reproductive Biotechnologies
GDARD	Gauteng Department of Agriculture and Rural Development
LLFPH	low inclusion of liquid fermented potato hash with and without enzyme, g/kg
HLFPH	high inclusion of liquid fermented potato hash with and without enzyme, g/kg
HOST	hypo-osmotic swelling test
HLFPH+E	high liquid fermented potato hash with enzyme diet, g/kg
LLFPH+E	low liquid fermented potato hash with enzyme diet, g/kg
LFC + E	Liquid fermented control with enzyme diet, g/kg
LFC	liquid fermented control diet, g/kg
LW × LR	Large White × Landrace
MED	medium, %
NPM	non progressive motility, %

SC	spermatozoa concentration, 106/mL
PSO	pomegranate seed oil
VSL	straight-line velocity, μm/s
STR	straightness, %
LIN	linearity, %
RPS	raw potato starch
LH	luteinizing Hormone
WOB	Wobble, %
TM	total motility, %
VAP	average path velocity, μm/s
PM	progressive motility, %
μL	microliter
VCL	curvilinear velocity, μm/s
%	percentages
PKC	palm kernel cake
°C	degree celscius
SLW	slow, %
RAP	rapid motility, %
STC	static, %

Author details

Thivhilaheli Richard Netshirovha[1], Zwivhuya Constance Raphalalani[2], Masindi Lottus Mphaphathi[3*], Mammikele Tsatsimpe[4], Lerato Deirdre Sehlabela[3] and Pulane Sebothoma[3]

1 College of Agriculture and Environmental Science, University of South Africa, Florida, South Africa

2 North West Department of Agriculture and Rural Development, Animal Science, Potchefstroom, South Africa

3 Agricultural Research Council, Animal Production, Germplasm Conservation and Reproductive Biotechnologies, Irene, South Africa

4 Gauteng Department of Agriculture and Rural Development, Johannesburg, South Africa

*Address all correspondence to: masindim@arc.agric.co.za

References

[1] Ogunlade JT, Ewuola EO, Gbore FA, Bandyopadhyay R, Niezen J, Egbunike GN. Testicular and epididymal sperm reserves of rabbits fed fumonisin contaminated diets. World Applied Sciences Journal. 2006;**1**(1):35-38

[2] Uno UU, Okolo CM, Ogbe HO, Ekaluo UB, Benjamin ME. Evaluation of spermatotoxic effect of sweet potato (Ipomoea batatas) leaf extract on male albino rats. Asian Journal of Biotechnology and Bioresource Technology. 2017;**1**(3):1-5

[3] Uno UU, Ekpo PB, Okolo CM, Ekaluo UB. Comparative effects of sweet potato (Ipomoea batatas) leaf and tuber on male albino rats. Asian Journal of Research in Medical and Pharmaceutical Sciences. 2017;**1**(2):1-7

[4] Ogbuewu IP, Omede AA, Chukwuka OK, Iheshiulor OOM, Uchegbu MC, Udebuani AC, et al. The overview of the chemistry, health benefits and the potential threats associated with prolonged exposure to dietary soy isoflavones. A review. International Journal Agriculture Research. 2010;**5**(12):1084-1099

[5] Claus R, Losel D, Lacorn M, Mentschel J, Schenkel H. Effects of butyrate on apoptosis in the pig colon and its consequences for skatole formation and tissue accumulation. Journal of Animal Science. 2003;**81**:239-248

[6] Andersson K, Schaub A, Andersson K, Lundström K, Thomke S, Hansson I. The effects of feeding system, lysine level and gilt contact on performance, skatole levels and economy of entire male pigs. Livestock Production Science. 1997;**51**:131-140

[7] Zamaratskaia G, Zlabek V, Ropstad E, Andresen Ø. In vitro and in vivo association of porcine hepatic cytochrome P450 3A and 2C activities with testicular steroids. Reproduction in Domestic Animals. 2012;**47**(6):891-898. DOI: 10.1111/j.1439-0531.2012.01986.x

[8] Netshirovha TR, Umesiobi DO, Matabane MB, Mphaphathi ML, Pilane CM, Thomas SR, et al. Improvement of Kolbroek boar growth performance and carcass quality through dietary crude protein supplementation. Open Journal of Animal Sciences. 2020;**10**:502513. DOI: 10.4236/ojas.2020.10303

[9] Netshirovha TR, Umesiobi DO, Matabane MB, Mphaphathi ML, Pilane CM, Thomas SR, et al. Improvement of Kolbroek boar growth performance and carcass quality through dietary crude protein supplementation. Open Journal of Animal Sciences. 2020;**10**:502-513. DOI: 10.4236/ojas.2020.103031

[10] Borg KE, Lunstra DD, Christenson RK. Semen characteristics, testicular size, and reproductive hormone concentrations in mature Duroc, Meishan, Fengjing, and Minzhu boars. Biology of Reproduction. 1993;**49**:515-521. DOI: 10.1095/biolreprod49.3.515

[11] Borg KE, Lunstra DD, Christenson RK. Semen characteristics, testicular size, and reproductive hormone concentrations in mature Duroc, Meishan, Fengjing, and Minzhu boars. Biology of Reproduction. 1993;**49**:515-521

[12] Rijnen MMJA, Dekker RA, Bakker GCM, Verstegen MWA,

Schrama JW. Effects of dietary fermentable carbohydrates on the empty weights of the gastrointestinal tract in growing pigs. The Digestive Physiology of Pigs Proceedings of the 8th Symposium. Uppsala, Sweden: Swedish University of Agricultural Sciences; 2001

[13] Canibe N, Jensen BB. Fermented and nonfermented liquid feed to growing pigs: Effect on aspects of gastrointestinal ecology and growth performance. Journal of Animal Science. 2019;**81**(8): 2019-2031. DOI: 10.2527/2003.8182019x

[14] Missotten JAM, Michiels J, Ovyn A, Desmet S, Dierick NA. Fermented liquid feed for pigs. Archives of Animal Nutrition. 2010;**64**:437-466

[15] Briz MDS, Bonet B, Pinart R. Spermatozoa malformations throughout the boar epididymis duct camps. Animal Reproduction Science. 1995;**43**(4):221-239. DOI: 10.1016/0378-4320(96)01470-4

[16] Thomas RS, Kanengoni AT, Chimonyo M. Impact of fermented liquid potato hash diets on growth performance of grower pigs. Journal of Agricultural Science. 2018;**10**:6

[17] National Research Council (NRC). Nutrient Requirements of Swine. 11th ed. Washington, DC, USA: National Academies Press; 2012

[18] Sehgal JP, Kumar BB. Supplementation of fermented yeast culture augments the growth and reduces the age at puberty in male Murrah buffalo calves. Buffalo Bulletin. 2016;**35**:179-190

[19] Lovercamp KW, Safranski TJ, Fischer KA, Manandhar G, Sutovsky M, Herring W, et al. High resolution light microscopic evaluation of boar semen quality sperm cytoplasmic droplet retention in relationship with boar fertility parameters. Archives of Andrology. 2007;**53**:219-228

[20] Gofura A, Witjoroa A, Ningtiyasa EW, Setyowatia E, Mukharromaha SA, Suhartinia Atho'illahb MF, et al. The ameliorative effect of black soybean and purple sweet potato to improve sperm quality through suppressing reactive oxygen species (ROS) in type 2 diabetes mellitus rat (Rattus novergicus). Science Asia. 2018;**44**:303-310

[21] Hyacinth AA, Abel OO, Terzungwe A, Daniel OL. Testicular and Epididymal characteristics of rabbit bucks fed *Tephrosia bracteolata* leaf meal. International Journal of Livestock Research. 2016;**6**(16):2277-1964

[22] Nur Z, Cakmak S, Ustuner B, Cakmak I, Erturk M, Abramson C, et al. The use of the hypo-osmotic swelling test, water test, and supravital staining in the evaluation of drone sperm. Apidologie. 2012;**43**:31-38. DOI: 10.1007/s13592-011-0073-1

[23] dos Santos AX, Kahwage PR, Faturi C, Quinzeiro NT, Lourenço J, Joele MRSP, et al. Feed supplementation with palm kernel cake-based concentrate increases the quality of water buffalo semen. Animal Reproduction. 2014;**11**(2):85-95

[24] Wähner M, Geyer M, Hallfarth G, Hühn V. Der einfluss von Zulagen einer Vitaminemulsion mit l-carnitine auf die Spermaeigenschaften von Besamungsebern. Züchtungskunde. 2004;**76**:196-207

[25] Hadi A, Abbass YI, Yadgar MA. The impact of L-carnitine supplement on semen variables and the levels of sexual hormones (serum LH, FSH, testosterone, and inhibin) in males with infertility. Medico-Legal Update.

2020;**20**(1):772-776. DOI: 10.37506/v20/i1/2020/mlu/194418

[26] Awoniyi DO, Aboua YG, Marnewick J, Brooks N. The effects of rooibos (Aspalathus linearis), green tea (Camellia sinensis) and commercial rooibos and green tea supplements on epididymal sperm in oxidative stress-induced rats. Phytotherapy Research. 2012;**26**:1231-1239

[27] Amao OA, Showunmi KA. Reproductive characteristics of rabbit bucks fed diet containing raw or fermented cottonseed cake. British Biotechnology Journal. 2016;**10**:1-7

[28] Chung F, Rubio J, Gonzales C, Gasco M, Gonzales G. Dose– Response effects of Lepidium meyenii (Maca) aqueous extract on testicular function and weight of different organs in adult rats. Journal of Ethnopharmacology. 2005;**98**:143-147

[29] Ekpo PB, Uno UU, Ogbe HO, Ekaluo UB. Effect of sweet potato (Ipomoea batatas) tuber on sperm profile and testicular integrity of male albino rats. Archives of Current Research International. 2017;**9**:1-7

[30] Sansone G, Nastri MJF, Fabbrocini A. Storage of buffalo (Bubalus bubalis) semen. Animal Reproduction Science. 2001;**62**:55-76

[31] Vale WG. Reproductive management of buffalo male aiming semen production for artificial insemination. In: Abstract of the 1st Buffalo Symposium of Americas. Belém, Pará, Brazil. Belém: PRODEPA; 2002. pp. 156-171

[32] Pant HC, Sharma RK, Patel SH, Shukla HR, Mittal AK, Kasiraj R, et al. Testicular development and its relationship to semen production in Murrah buffalo bulls. Theriogenology. 2003;**60**:27-34. DOI: 10.1016/S0093-691X(02)01037-3

[33] KA, Egbunike GN, Woogeng IN. Evaluation of the fertility of broiler breeder cocks fed on diets containing differently processed sweet potato tuber in a humid tropical environment. Global Journal of Poultry Farming and Vaccination. 2013;**4**(5):82-87

[34] Pant HC, Sharma RK, Patel SH, Shukla HR, Mittal AK, Kasiraj R, et al. Testicular development and its relationship to semen production in Murrah buffalo bulls. Theriogenology. 2003;**60**:27-34

[35] Huang YT, Johnson RK. Effect of selection for size of testes in boars on semen and testis traits. Journal of Animal Science. 1996;**74**(4):750-760. DOI: 10.2527/1996.744750x

[36] Zhao J, Jin Y, Du M, Liu W, Ren Y, Zhang C, et al. The effect of dietary grape pomace supplementation on epididymal sperm quality and testicular antioxidant ability in ram lambs. Theriogenology. 2017;**97**:50-56

[37] McDaniel CD, Hannah JL, Parker HM, Smith TW, Schultz CD, Zumwalt CD. Use of a sperm analyser for evaluating broiler breeder males. Effects of altering sperm quality and quantity on the sperm motility index. Poultry Science. 1998;**77**:888-893

[38] dos Santos MN, Ramachandran R, Kiess AS, Wamsley KGS, McDaniel CD. The impact of dietary yeast fermentation product derived from Saccharomyces cerevisiae on semen quality and semen microbiota of aged white Leghorn roosters. Journal of Applied Poultry, Research. 2018;**27**:488-498

[39] Van Dorland A, Bruckmaier R, Wach-Gygax L, Jeannerat E, Janett F,

Sieme H, et al. Variability of antioxidant capacity of stallion semen during feed supplementation with a yeast product. Journal of Equine Veterinary Science. 2016;**43**:81-82

[40] Abaza IM, Shehata MA, Shoieb MS. Evaluation of some natural feed additive in layer diets. Poultry Science. 2006;**26**:891-909

[41] Ragab A, Zohdi W, Awad H, Azab S, Salem H, Elkaraksy A, et al. Assessment of human sperm morphology: Comparison ofstrict Kruger's criteria versus inverted microscopy motile spermorganelle morphology examination (MSOME). European Urology Supplements. 2017;**15**:e814

[42] Lovercamp KW, Safranski TJ, Fischer KA, Manandhar G, Sutovsky M, Herring W, et al. High resolution light microscopic evaluation of boar semen quality sperm cytoplasmic droplet retention in relationship with boar fertility parameters. Archives of Andrology. 2007;**53**:219-228. DOI: 10.1080/01485010701426463

[43] Bréque C, Surai P, Brillard JP. Roles of antioxidants on prolonged storage of avian spermatozoa in vivo and in vitro. Molecular Reproduction and Development. 2003;**66**:314-323

[44] Njoku R-CC, Abarikwu SO, Uwakwe AA, Mgbudom-Okah CJ, Ezir CY. Dietary fluted pumpkin seeds induce reversible oligospermia and androgen insufficiency in adult rats. Systems Biology in Reproductive Medicine. 2019;**65**:437-450

[45] Skoracka K, Eder P, Łykowska-Szuber L, Dobrowolska A, Krela-Kaźmierczak I. Diet and nutritional factors in male (In)fertility-underestimated factors. Journal of Clinical Medicine. 2020;**9**(5):1400. DOI: 10.3390/jcm9051400

[46] Ogbuewu IP, Okoli IC, Iloeje MU. Semen quality characteristics, reaction time, testis weight and seminiferous tubule diameter of buck rabbits fed neem (Azadirachta indica a. Juss) leaf meal-based diets. Iranian Journal of Reproductive Medicine. 2009;**7**:23-28

[47] Ansa AA, Akpere O, Imasuen JA. Semen traits, testicular morphometry and histopathology of cadmium-exposed rabbit bucks administered methanolic extract of Phoenix dactylifera fruit. Maringá. 2017;**39**(2):207-215

[48] Perry G, Petterson D. Determining reproductive fertility in herd bulls. University of Missouri Agriculture publication. Mu extension, University of Missouri-Columbia. 2001:1-8

[49] Adienbo OM, Wodu CO. Fertility-enhancing potentials of ipomoea batatas (sweet potato) in male albino Wistar rats. European Journal of Biomedical and Pharmaceutical Sciences. 2015;**2**(3):1197-1202

[50] Bovula N, Ncobela CN, Pilane CM, Nedambale TL, Chimonyo M. Growth performance and fertility of Windsnyer boars supplemented with α-tocopherol. Tropical Animal Health and Production. 2021;**53**:161. DOI: 10.1007/s11250-021-02610-9

[51] Handelsman DJ. Androgen Physiology, Pharmacology, Use and Misuse. South Dartmouth, MA: National Library of Medicine; 2020

[52] Bitto II, Sende CT, Eze PU. Preliminary investigation on the effect of cassava peel meal on testicular morphometric and some biochemical characteristics of serum in cockerels.

Global Journal of Pure and Applied Sciences. 1999;**6**(2):161-165

[53] Majid M, Ijaz Fatima M, Baig W, Nasir B, Khan MR, Haq I. Scientific validation of Ethnomedicinal use of Ipomoea batatas L. lam. as aphrodisiac and Gonadoprotective agent against bisphenol a induced testicular toxicity in male Sprague Dawley rats. Hindawi BioMed Research International. 2019:1-21. Article ID 8939854. DOI: 10.1155/2019/8939854

Chapter 8

Using Conventional Ruminant Techniques and Molecular Spectroscopy to Study the Impact of Additive Fibrolytic Enzymes and Maturity Stage on Nutritional and Molecular Structural Changes of Legume and Legume-Cereal Intercropped Silage

Victor Guevara, Carlene Nagy, Jen-Chieh Yang, Jiangfeng He, Maria E. Rodriguez-Espinosa, Weixian Zhang, Tao Ran and Peiqiang Yu

Abstract

This chapter aims to I) provide research background and motivation on the impact of additive fibrolytic enzyme and maturity stage at harvesting on molecular structural changes and nutritional value of the cool-season legume silage and legume-cereal intercropped silage; II) provide recent research progress and development in whole plant faba bean (legume) silage and faba-oat (legume-cereal) intercropped silage. The reviewed projects include: I) effect of adding different levels of additive fibrolytic enzymes on utilisation of cool-season whole plant faba bean silage in ruminants to find an optimal dose level for this faba silage; II) effect of adding different levels of fibrolytic enzymes on utilisation of cool-season intercropped whole plant faba-oat (legume-cereal) silage in ruminants; III) effect of maturity stage at harvesting on nutritive quality of whole plant faba silage; IV) effect of frost damage on nutritive quality of whole plant faba forage in ruminant; V) feeding trial and dairy production performance, milk yield (ECM, FCM, fat yield etc.) with whole plant faba legume silage in early lactating cows to replace traditional barley and corn silages; VI) availability and utilisation of whole plant faba silage and intercropped whole plant faba-oat intercropped silage in ruminants; VII) using molecular spectroscopy to study nutrition and structure interaction of faba silage at cellular and molecular levels. Based on the scientific findings presented in this chapter, the following most important conclusions can be drawn: cool-season faba (legume) variety with different tannin levels impact not only nutrient profiles but also protein and carbohydrate-related molecular structure makeup. Additionally, the nutrient supply, bioenergy, degradation, digestion, and metabolic characteristics of cool-season faba silage and intercropped faba-oat silage were highly related independently and synergistically to molecular structure conformation. Furthermore, the nutrient utilisation and availability of cool-season faba silage and intercropped silage in ruminant livestock systems could be accurately predicted by the protein and carbohydrate molecular structures revealed with cutting stage vibrational molecular spectroscopy when they work together. Additive fibrolytic enzyme and maturity stage at harvesting significantly impacted both nutritional and molecular structural changes of legume and legume-cereal intercropped silage. Dairy production performance and milk yield (ECM, FCM, fat yield, etc.) studies showed that whole plant faba legume silage in early lactating cows could be used as an alternative silage to traditional barley and corn silages. The information described in this chapter gives better insight into cool-season legume silage and legume-cereal intercropping silage research progress in terms of inherent molecular structures, nutritive quality, animal production response, and molecular structure and nutrition delivery interactive relationship as well as impact by maturity stage and dosage levels of additive fibrolytic enzymes in the cool-season legume silage and intercropped legume-cereal silages.

Keywords: feed additive, whole plant legume silage, intercropping legume-cereal silage, fibrolytic enzyme, vibrational spectroscopy, molecular structure, nutrient utilisation and availability, ruminant systems, molecular structure- nutrition delivery interaction, animal production

1. Introduction

As new cool-season faba bean varieties (high-tannin, low-tannin, and zero-tannin) are developed and available in western Canada and production has been increasing in recent years [1–4], utilisation of this faba legume as forage hay or silage is possible. To our knowledge, no systematic study on the nutrition quality of these cool-season whole crop faba beans as hay and silage has been found, and there is no study on whole crop faba and whole crop faba-oat intercropping silages from other research teams. Also, there is no research on the impact of maturity cutting stages: flower stage, mid-pod stage, and late-pod stage, on the feed nutritive quality of whole crop faba (legume) as hay and silage, as well as whole crop faba bean-oat intercropping forage (as silage or as hay) in dairy, sheep, goat, and beef cattle (all type of ruminants).

Recently, innovative mixed fibrolytic enzymes (FE) have been used to improve feed nutritive value and utilisation in dairy and beef cattle by increasing polysaccharide or feed fibre degradability in the rumen and digestibility in the whole gastrointestinal tract [5, 6]. These innovative fibrolytic enzymes are able to release trapped polysaccharides after breaking down chemical functional group bonds (e.g. ester bonds and ether bonds) between lignin and polysaccharides through hydroxycinnamic acid and ferulic acid bridges in the complex plant cell wall. However, there is no study on the optimal dosage of these innovative enzymes on whole crop faba (legume) silage and whole crop faba-oat (legume-cereal) intercropping silage in the literature [7].

Advanced synchrotron-radiation and Globar-sourced vibrational (micro)spectroscopy is capable of revealing internal structure features at cellular and molecular levels and simultaneously provides four kinds of important information: chemical composition, molecular structure, environment, and chemistry, within intact tissue with a highly spatial resolution [8–15]. However, to date, these advanced vibration molecular techniques are still seldomly known to animal and feed scientists, particularly synchrotron technology. There is no study on using these advanced molecular spectroscopic techniques in legume silage (e.g. whole crop faba bean silage) and legume-cereal intercropped silages (e.g. whole crop faba bean-oat silage) in literature from other research groups.

The objective of this chapter is to provide research background and motivation on impact of additive fibrolytic enzymes and maturity stage at harvesting on molecular structural changes and nutritional value of the cool-season legume silage and legume-cereal intercropped silage; provide recent research progress and development in whole plant faba bean (legume) silage and faba-oat (legume-cereal) intercropped silage in ruminant system [4, 7, 16]. The information described in this chapter gives better insight into legume silage and legume-cereal intercropping silage research progress in molecular structure, nutrition and molecular structure interactive relationship, and animal production response to these cool-season legume silage or cool-season legume-cereal silage.

2. Recent research and progress in cool-season whole crop faba legume silage

2.1 Recent study on the effect of cutting stage and tannin content on nutritive quality of cool-season whole crop faba legume silage

Information regarding the utilisation of the cool-season whole crop faba legume silage in beef and dairy cattle is extremely limited [4]. Our team member, Guevara [4] carried out systematic studies to reveal the impact of faba legume tannin contents and newly developed cool-season varieties from Crop Development Centre (CDC, University of Saskatchewan) (Snowdrop variety with low-tannin content; SSNS-1 genotype with high-tannin content) and the impact of maturity cutting stages (at 88-days faba legume mid-pod stage; at 97-days faba legume late pod stage) on faba legume forage yield, nutrient profiles, bio-energy content (TDN, ME, and NE value) [17, 18], protein and carbohydrate subfractions and nutrient supply evaluated with the Cornell Net Carbohydrate and Protein System (CNCPS) [19, 20], rumen fermentation kinetics [21–23], potential rumen available nitrogen (ED_N) to rumen available energy (ED_E) synchronisation and degradation ratios [24, 25], intestinal digestibility of primary nutrients [26], metabolic characteristics [17, 22], and predict production performance in term of feed milk value (FMV) when utilisation of cool-season whole crop faba legume silage in lactation dairy cattle.

Our team member, Guevara [4] found that the yield on a dry matter (DM) basis of cool-season whole crop faba legume silage cut at the faba flower stage was lower than that at the late-pod stage (7.34 vs. 12.20 tons per ha). Guevara also found that there was much higher in the rumen pH, ammonia (NH_3) production, and volatile fatty acid (VFA) in terms of rumen acetic acid and propionic acid at the faba flower stage in the cool-season whole crop faba legume silage than that at the faba legume mid-pod stage and late-pod stage (5.39 vs. 4.35 and 4.51; 16.32 vs. 6.62 and 5.66% of total N; 6.33 vs. 2.35 and 1.70% DM; 1.44 vs. 0.04 and 0.06% DM, respectively). There was no significant difference in crude protein (CP) content among different cool-season silage varieties and among all three different maturity stages (CP: 22% of DM). However, there was no difference in net energy of lactation (NE_{L3x}) for dairy cows in cool-season silage when cut faba legume was at mid-pod and late-pod stages, but they were higher than cool-season faba legume silage when cut at flower stage (1.45 and 1.46 vs. 1.13 Mcal/kg DM). Starch content in cool-season faba legume silage was higher when cut at faba legume late-pod stage than that cut at faba legume mid-pod and flower stages (17.2 vs. 9.4 and 1.3% of DM). For fibre content in the cool-season faba legume silages, there was no difference between mid-pod and late-pod stages, but much lower when cut at faba flower stage (36.0 and 34.4 vs. 45.3% DM). The rumen degradation study by Guevara [4] showed that the rumen undegraded protein (RUP^{NRC}) was higher when cut at faba late-pod stage and rumen undegraded/bypass starch (BSt) was higher when cut at faba mid-pod stage (33 vs. 25 and 32 vs. 18 g/kg DM, respectively).

The results from Guevara's [4] study in N to energy synchronisation showed a lower rumen available N to rumen available carbohydrates (ED_N/ED_CHO) overall ratio when cutting faba legume at the late-pod stage in comparison with faba silage cut at the mid-pod stage (−35 g/kg). The intestinal absorbable faba protein (IADP) and the total tract digested faba starch (TDST) were higher (84 vs. 61 g/kg CP and 175 vs. 95 g/kg DM, respectively) when cut at late-pod stage. Both the DEV/OEB and NRC protein systems showed that there was a lower in total truly digestible protein supply (DVE value: 59 vs. 68 g/kg DM), total metabolizable protein (MP: 67 vs. 73 g/kg DM) and feed milk value based on the DVE (1.20 vs. 1.37 kg milk per kg DM faba silage) or MP value (1.36 vs. 1.48 kg milk per kg DM faba silage) when faba legume silage was cut at mid-pod stage than that when cut at late-pod stage.

Then Guevara [4] concluded that in order to obtain high yield and high feed nutritional values, the cool-season faba legume forage should be harvested at the late-pod stage. In this late-pod stage, cool-season faba legume silage showed high predicted production performance. Therefore, the cool-season whole crop faba legume silage when cut at the late pod stage has the highest nutritional value and greatest potential to be used as an alternative ingredient in dairy and beef rations.

2.2 Recent study on the effect of frost damage on feed quality of cool-season whole crop faba legume forage for ruminants

Frost-damaged faba bean plants often happen due to the cold weather in western Canada. What nutritional value and how to utilise this frost-damaged faba bean is a question. Therefore, recently, our team, Guevara [4] systematically evaluated the impact of faba silage tannin contents (high level vs. low level vs. zero) and cool-season faba cultivars (CDC developed Snowdrop variety with lower level tannin content; CDC SSNS-1 variety with a high-tannin content) on physiochemical nutrient profiles, bioenergy value (TDN, ME, NE) for dairy and beef cattle [17, 18], faba protein and carbohydrate CNCPS subfractions and CNCPS nutrient supply [19, 20], rumen fermentation and degradation kinetics of rate and extent [21–23], potential rumen available N to rumen available energy synchronisation and hourly effective degradation ratios [24, 25], intestinal digestion of nutrients [26], metabolic characteristics (e.g. DVE, OEB, and MP values) [17, 22], and predicted animal production performance in terms of feed milk value of the cool-season frost-damaged whole crop faba legume hay harvested at 114 days of maturity stage [4].

In these studies, Guevara [4] reported that compared with cool-season low-tannin frost-damaged whole crop faba legume hay, the high-tannin frost-damaged faba legume hay was higher in organic matter and lower in acid detergent insoluble crude protein (ADICP: +2.5%DM and − 0.4%DM). However, there was no difference in starch and crude protein (CP) content at this maturity stage between the high- and low-tannin varieties with an average of 11.9% DM and 16.8% DM, respectively. The bio-energy values in terms of TDN, ME, and NE for dairy and beef cattle were also not different. However, there was a higher in fibre-bound protein (PB2) and lower indigestible protein (PC, +2.3 and − 3.1% CP) in the high-tannin frost-damaged faba hay than that in the low-tannin frost-damaged faba legume hay.

Rumen kinetic study [4] showed that the cool-season frost-damaged low lignin faba hay was higher in rumen bypass or undegraded protein with a RUP of +2.8% and lower in rumen undegradable neutral detergent fibre (NDF) fraction (U, −5.7%) compared to the frost-damaged high lignin faba hay. The intestinal phase study [4] showed that there was higher in various nutrient supplies and predicted production performance from the cool-season frost-damaged high lignin faba hay than that in the cool-season low lignin frost-damaged faba hay in terms of intestinal digested rumen undegraded DM (IDBDM, +15 g/kg DM), total metabolizable protein (MP, +4 g/kg DM), intestinal digestibility of rumen bypass or undegraded feed protein (dIDP, +7%), and feed milk value (FMV^{NRC}, +0.09 kg milk per kg DM faba hay).

Then Guevara [4] concluded that compared with the non-frost damage cool-season faba hay [4], both frost-damaged cool-season high- and low-tannin faba hay were lower in feed quality and nutritional supply at 114 days than the non-frost damage cool-season faba hay when cut at the faba flower stage (77 days), faba mid-pod stage (88 days), and faba late-pod stage (97 days). However, within the frost-damaged cool-season faba forage, the cool-season frost-damaged high-tannin faba hay which was harvested at a growth stage of 114 days had superior feed quality and nutritional value than that in the cool-season frost-damaged low-tannin faba hay which was harvested at the growth stage of 114 days in western Canada.

2.3 Recent studies in feeding trial and dairy production performance and metabolic characteristics with cool-season whole crop faba legume silage in high producing cows to replace conventional barley and corn silage

How to feed the cool-season whole crop faba legume silage and what animal production performance in comparison with conventional barley and corn silage in high lactating dairy cows are still not known from the literature. Therefore, our team [4, 27] conducted a dairy trial experiment to determine the impact of 50% and 75% partial silage replacements (T50, T75) in dairy cows' rations and 100% complete silage replacement (T100) in dairy cows' rations containing conventional corn and barley silages with the cool-season low-tannin Snowdrop variety of whole crop faba legume silage cut at faba late-pod stage (97 days old) on early lactating dairy cows (high production) in terms of dairy production performance of milk yield and components, DM feed intake and feed-milk efficiency, N balance, intestinal digestibility, rumen degradation and fermentation features, and metabolic characteristics as well as dairy cow feeding behaviour. This experiment that we used was a double 4 × 4 Latin square design (LSD) with four non-cannulated and four cannulated lactating cows). Each period of LSD lasted 25 days, including adaption and sampling collection.

Guevara et al. [27] and Guevara [4] reported that our results showed that the dairy cows fed T100 with 30.60% cool-season whole crop faba legume silage produced higher fat corrected milk yield (3.5% FCM) and higher energy corrected milk yield (ECM) than the cows fed a control diet of T0 with 18.37% corn silage +12.23% barley silage (+4.35 and + 3.48 kg/cow/d, respectively), but there was no significant difference in FCM and ECM when lactating dairy cows were fed T50 with 9.18% corn silage

+6.12% barley silage +15.30% cool-season whole crop faba legume silage and fed T75 with 4.59% corn silage +3.06% barley silage +22.95% cool-season whole crop faba legume silage.

Our results also showed that when lactating dairy cows fed diets containing cool-season whole crop faba legume silage (T50, T75, T100) in comparison with control T0 produced higher milk fat yield (2.11 vs. 1.89 kg per cow per day). A feed efficiency study [4, 27] showed when lactating dairy cows consumed T75, FCM/DMI was higher than the lactating cows when consumed control T0 diet (2.21 vs. 1.91). There was no difference in starch digestibility of lactating dairy cows among the three cool-season faba silage-containing diets: T50, T75, and T100, but they were all lower than control T0 diet without any faba legume silage (92.65% vs. 96.13%).

Our dietary energy study results showed that the cow diets contained or included the cool-season whole crop faba legume silage (T50, T75, T100 diets vs. control T0 diet) significantly increased the total diet energy (1.91 vs. 1.65 Mcal per kg DMI), percentage of energy for cow body weight gain and total milk production (78.3 vs. 75.5% of total energy). The study showed similar rumen fermentation features in ammonia, VFA, and pH) among all the treatment diets (T0, T50, T75, and T100).

Then our team, Guevara et al. [27] and Guevara [4], concluded that the dietary inclusion of cool-season whole crop faba legume silage, which was cut at faba late-pod stage improves both fat and energy corrected milk yield, and also increases milk fat yield, and improves efficiency (FCM/DMI) without negatively affecting the DM intake. Consequently, our study [4, 27] showed that the cool-season whole crop faba legume silage cut at faba late-pod stage is a highly nutritive alternative feed which can improve dairy cow production performance in western Canada. In this feeding trail, we, Guevara et al. [27] and Guevara [4], also studied cost–benefit in terms of the income over feed cost (IOFC). The results showed a superior benefit to dairy farmers when using cool-season whole crop faba legume silage cut at late-pod stage to replace conventional barley and corn silages in high producing dairy cows [4].

3. Recent research and progress in cool-season whole crop faba and whole crop oat intercropping hay and silage

3.1 Recent study on the effect of maturity stage/cutting time on yield, chemical, and nutrient profiles, predicted production performance of cool-season whole crop faba and whole crop oat intercropping legume-cereal hay

The high protein and high starch content in whole crop faba legume forage make them suitable for ruminant diets. There is very limited information regarding the utilisation of cool-season whole crop faba legume hay for ruminants [16]. Therefore, our team member, Nagy [16], conducted experiments to study the impact of intercropping cool-season whole crop oat-faba for hay and the impact of the cutting stage on yield, chemical composition, and bio-energy profile [17, 18], protein and carbohydrate CNCPS fractions and CNCPS nutrient supply [19, 20], rumen degradation kinetics [21–23], N to energy synchronisation and degradation ratios [24, 25], intestinal digestibility [26], metabolic characteristics (e.g. MP, DVE, and OEB) [17, 22], and predicted dairy production performance of cool-season whole crop oat-faba (legume-cereal) hay. In our study [16], the oat and faba plant were intercropped and grown in three fields in Saskatchewan, Canada, and were cut at three growth stages for hay and silages: Cutting stage 1 with the oat plants at the inflorescence stage and the faba bean plants at the flat pod stage; Cutting stage 2 with the oat plants at the milk development stage and the faba bean plants at the milk pod stage; Cutting stage 3 with oat plants at the soft dough stage and the faba bean plants at the late pod stage.

The chemical compositions of cool-season whole crop faba-oat hay were determined using standard chemical analysis methods [28]. Bio-energy values and total digestible nutrients (TDN) were evaluated using the NRC chemical approach [17, 18], protein and carbohydrate subfractions and CNCPS nutrient supply [19, 20] were determined using the updated CNCPS 6.5 system. The rumen degradation was carried out using a standard *in situ* technique [21] with rumen cannulated lactating cows at our dairy research facility (RDTRF, Saskatoon, University of Saskatchewan, Canada). The rumen available N to rumen available energy potential synchronisation and hourly effective ED_N to ED_OM ratios were evaluated using Tas et al.'s method [24] developed by Wageningen University and Research, The Netherlands. The intestinal digestion was evaluated using the modified three-step *in situ* and *in vitro* method with pre-*in situ* 12 h incubation [26]. The truly digestible protein supply (DVE), protein degraded balance (OEB), net energy-based FMV, and metabolizable protein-based FMV [22, 23] were evaluated using both updated DVE/OEB and NRC nutritional systems.

Nagy [16] showed that cutting stages 2 and 3 had a higher DM hay yield than cutting stage 1. With increasing cutting stage, ash and soluble protein (SCP) were decreased from 14.1 to 9.6% DM and 13.3 to 9.8% DM, respectively. With increasing cutting stage, the starch, sugar, and non-fibre carbohydrate (NFC) contents in cool-season intercropped faba-oat hay were dramatically increased from 0.31 to 7.1% DM (starch), 6.6 to 13.0% DM (sugar), and 15.3 to 24.7% DM (NFC). The stage of cutting did not significantly impact NDF, acid detergent fibre (ADF), and acid detergent lignin (ADL). There was no difference among the three cutting stages in the cool-season whole crop faba-oat hay. In the TDN and bioenergy studies, the results showed that with extending cutting stage, the tdNFC, TDN value, NE for lactation, NE for growth, and NE for maintenance in both beef and dairy cattle increased from 12.7 to 22.1% DM, 49.1 to 56.0%DM (TDN), 1.01 to 1.18 Mcal/kg (NE_{L3x}), 0.51 to 0.69 Mcal/kg (NE_g), and 1.07 to 1.26 Mcal/kg (NE_m), respectively. These results suggest that the cool-season whole crop faba-oat (legume-cereal) hay can be used as a high-quality forage for both beef and dairy cattle in western Canada.

3.2 Recent study on the effect of maturity stage/cutting time on silage yield, chemical profile, energy and protein-based feed milk value and metabolic characteristics of cool-season whole crop faba-oat intercropping legume-cereal silage

Recently, Nagy [16] conducted a systematic study on the impact of maturity cutting stage and the intercropping of cool-season whole crop faba with whole crop oat legume-cereal silage on intercropped silage yield, nutritive value profiles, protein and carbohydrate CNCPS subfractions and CNCPS nutrient supply, bio-energy content, ruminal fermentation kinetics features, rumen available N to rumen available energy synchronisation and hourly effective degradation ratio, intestinal digestion, and truly absorbable protein supply in term of DVE and MP values to dairy cows. The cool-season CDC oat and faba bean were intercropped, grown in three fields, and were cut at three growth stages: Cutting stage 1 with the oat plants at the inflorescence stage and the faba plants at the flat-pod stage; Cutting stage 2 with the oat plants at the milk development stage and the faba plants at the mid-pod stage; Cutting stage 3 with oat plants at the soft dough stage and the faba plants at the late-pod stage. The chemical composition was determined using standard chemical analysis methods (e.g. AOAC), bioenergy values and TDN and its components (e.g. tdNDF, tdCP, tdFA, and tdNFC) were estimated using the NRC summary method, and protein and carbohydrate subfractions were determined using the updated CNCPS 6.5 system. The *in situ* techniques were used to determine rumen fermentation/degradation kinetic profiles with rumen cannulated lactating cows at our Rayner dairy station (RDTRF) at the University of Saskatchewan, Saskatoon, Canada. The rumen available N to rumen available energy synchronisation in terms of hourly effective degradation ratio - ED_N to ED_OM was determined using a method reported by Tas et al. [24] from Wageningen University and Research. The three-step *in situ* and *in vitro* method with pre-*in situ* incubation 12 h was applied to determine intestinal digestion of primary nutrients. The DVE/OEB system [22, 23] and NRC nutritional model [17] were used to determine the total truly digestible protein supply in the intestine and metabolizable protein (DVE, OEB, MP, etc.) and net energy-based FMV as well as metabolizable protein or DVE based FMV.

Nagy [16] reported that the cool-season whole crop faba-oat silage is higher in protein content in the 2nd and 3rd cutting growth stages than in the 1st cutting growth stage (20, 18 vs. 16% DM). The 3rd cutting stage had the highest TDN value (58 vs. 55, 48% DM). Additionally, the total MP and FMV^{NRC} were higher in the 2nd and 3rd cutting growth stages compared with the 1st cutting growth stage (MP: 65, 68 vs. 61 g/kg DM; $FMV^{NRC:}$ 1.31, 1.38 vs. 1.23 kg milk per kg of intercropped silage, respectively). These studies suggest that cutting stages 2 and 3 of cool-season intercropped faba-oat silage resulted in higher nutritive values (TDN, MP) and better predicted production performance.

4. Recent research and progress in feed additive impact on cool-season whole crop faba legume silage and cool-season whole faba-oat (legume-cereal) intercropping silage

4.1 Recent study on the impact of adding innovative fibrolytic enzyme (FE) at different dose levels on short-term and long-term degradability of cool-season whole crop faba legume silage in ruminant systems

Fibrolytic enzymes (FE) can be used to improve nutrient availability in ruminants by releasing cell-wall trapped nutrients in the complex plant cell wall and increasing fibre degradability and digestibility in animals [5, 6]. However, our literature research shows positive and no-effective impacts on dairy cows [29–34]. These results are due to several impacts such as dosage level, types of enzymes, conditions, diets, etc.

Recently, an innovative mixture of fibrolytic enzymes has been developed and it is able to release polysaccharides from complex cell walls after breaking down chemical bonds between lignin and polysaccharides [35]. Our team members, recently Yang et al. [36] and Yang [7] conducted several experiments to study the impact of adding the innovative fibrolytic enzyme (FE) at different dose levels on DM and NDF fibre degradability of the cool-season whole crop faba legume silage (cv. CDC Snowbird). We used both the Daisy^{II} *in vitro* incubation method and *in situ* nylon bag method to evaluate the degradability of DM (DMD) and neutral detergent fibre (NDFD), and we also compared these two different methods in evaluating the *in vitro* degradability.

In our experiments [7, 36], seven doses of innovative fibrolytic enzymes (IFE) were applied to the cool-season whole crop faba silage samples, including 0 (as control), 0.25, 0.5, 0.75, 1, 1.25, and 1.5 mL of FETR per kg DM of cool-season faba silage. Yang et al. [36] and Yang [7] reported that with increasing enzyme dosage levels, DMD was cubically impacted and NDFD was quadratically affected by the innovative fibrolytic enzymes in our *in situ* animal experiment. In the *in vitro* study, the dosage level quadratically affected DM degradability and cubically affected NDFD. When comparing the two different methods (*in vitro* vs. *in situ*), it was found that there existed a strongly or satisfactory correlationship between *in situ* and *in vitro* methods with r = 0.98 for overall DMD and r = 0.84 for overall NDFD.

Both our *in vitro* and *in situ* results showed that the DMD and NDFD were greatly impacted by this innovative fibrolytic enzyme in the cool-season whole crop faba legume silage. Although the Daisy^{II} *in vitro* technique showed some inconsistent and had a relatively larger variation when compared with the *in situ* nylon bag technique, it remains a rapid and useful tool to evaluate a large amount of samples or treatments in DM and neutral detergent fibre degradability with less cost, time, and labour.

4.2 Recent study on the effect of adding innovative fibrolytic enzyme (FE) at different dose levels on rumen fermentation characteristics and degradation kinetics of cool-season whole crop faba legume silage in ruminants

In this experiment, our team member, Yang [7] carried out an *in situ* animal trial to determine the impact of innovative fibrolytic enzymes (FETR) on DM and NDF fibre rumen fermentation and degradation kinetic characteristics of cool-season whole crop faba legume silage that we developed recently. In our study [7], the *in situ* animal trial was performed using two rumen cannulated Holstein cows in our dairy research station (RDTRF, the University of Saskatchewan, Canada) and the cool-season whole crop faba legume silage samples were treated with seven dosages of the innovative enzyme, including 0 (as control), 0.25, 0.5, 0.75, 1, 1.25, and 1.5 mL of FETR per kg DM of faba legume silage. *In situ,* rumen degradation residues and fermentation and degradation kinetics were determined.

Yang [7] reported that the innovative fibrolytic enzyme application linearly decreased DM degradation residue at 0 hour. Significant quadratic effects were observed at 3 hours (short-term) and 24 hours (long-term) of incubation. However, no significant differences in rumen degradation residues were found at other incubation time points. The rumen NDF degradation residue at 0, 6, and 24 hours was quadratically affected with the innovative fybrolytic enzyme addition, and a cubic impact was observed at 48-hour rumen incubation.

With increasing dosage levels of innovative fibrolytic enzyme, the rumen soluble fraction of DM (S_DM) was increased linearly in dairy cows from cool-season whole crop faba legume silage. The NDF rumen degradation kinetics were greatly affected by the innovative enzyme application by increasing the potentially degradable fraction (D_NDF) and effective degradable fibre content (ED_NDF) and reducing the undegradable fraction (U_NDF). Increasing dosage levels linearly increased the sum of washable and degradable (W + D) fractions and, therefore, linearly decreased the undegradable fraction. The dosage level of innovative enzyme also cubically impacted both rumen bypass NDF (BNDF) and effective degradable NDF (EDNDF).

Our results [7] indicated that the innovative fibrolytic enzyme significantly improved fibre fermentation and degradation for the cool-season whole crop faba legume silage. Yang [7] also suggested a further study in the near future to evaluate the impact of pre-treatment of the innovative fibrolytic enzyme derived from *Trichoderma reesei* on animal production performance (lactation), feeding behaviour, rumen function and metabolic parameters, intestinal and total tract digestibility in highly lactating dairy cows fed the cool-season whole crop faba legume silage as a main source of forage in comparison with conventional barley and corn silage.

4.3 Recent study on the effect of innovative fibrolytic enzyme (FE) at different dose levels on nutrient utilisation of cool-season whole crop faba bean-oat intercropping (legume-cereal) silage in ruminants

Recently, Nagy [16] conducted a study to analyse the impact of dosage level of innovative fibrolytic enzyme derived from *Trichoderma Reesei* (FETR) on *in vitro* fermentation and degradation kinetic features of intercropped cool-season whole crop oat-faba silage using rumen cannulated dairy cows. The cool-season CDC oat and CDC faba were intercropped and grown in three fields in Saskatchewan, Canada and were cut at the maturity stage with the whole crop oat at the soft dough stage and the whole crop faba at the late-pod stage of maturity. The degradation kinetics of primary nutrients were estimated using an *in vitro* technique with rumen liquid from rumen-fistulated lactating dairy cows. The *in vitro* rumen fermentation features and degradation kinetic characteristics of DM and fibres (both NDF and ADF), including rumen degradation rate (Kd), lag time (T0), potentially degradable fraction (D), rumen undegradable fraction (U), and rumen effective degradable fractions and content (ED) were evaluated. The treatment design was a one-way structure with 5 dosage levels of innovated fibrolytic enzyme (FETR: 0, 0.075, 0.15, 0.225, and 0.3 ml per litre). The experimental design was a RCBD with the dosage level as a fixed effect and animals and *in vitro* run as random block effects. The *in vitro* data were analysed using the mixed model procedure in SAS 9.4 with the analysis RCBD model. The orthogonal polynomial contrast (OPC) of SAS was used to study the relationship between dosage levels and *in vitro* degradation kinetics.

Nagy [16] reported that there were strongly significant interaction effects between enzyme dosage levels and incubation time for the degradability of DM (DDM), degradability of neutral detergent fibre (DNDF), and degradability of ADF (DADF). There was a cubic relationship between enzyme dosage levels and DDM ($P = 0.02$), a tended linear relationship with DNDF ($P = 0.06$), and a quadratic relationship with DADF ($P = 0.04$). The results [16] indicated that the dosage level of innovative fibrolytic enzyme and incubation time had a significantly synergistic effect on *in vitro* degradability of DM, NDF, and ADF in this intercropped cool-season whole crop oat-faba (legume-cereal) silage.

4.4 Recent study on the effect of fibrolytic enzyme (FE, exogenous) on lactational performance, milk yield (ECM, FCM, fat yield, etc), feeding behaviour, rumen fermentation and digestibility in lactating cows fed cool-season whole crop faba legume silage-based diet

In this study, our team members, Yang et al. [37] and Yang [7] also carried out studies to determine the impact of pre-treating cool-season whole crop faba legume silage based-diet with exogenous innovative fibrolytic enzyme derived from *Trichoderma reesei* (FETR) on animal production performance (lactation), intestinal and total tract digestibility, rumen degradation and fermentation features, energy partitioning, N balance, as well as eating behaviour in lactational dairy cows. This experiment was conducted with eight lactating Holstein dairy cows (body weight: 710 ± 44 kg; days in milk: 121 ± 17 days) with four different innovative enzyme dosage treatments, including 0 (as a control), 0.5, 0.75, and 1.0 ml of FETR per kg DM of cool-season faba legume silage diet) in a double 4 × 4 Latin square design (2-LSD). The innovative enzyme dosage applied to cool-season faba silage diet in this experiment was selected based on our previous studies. They showed a positive impact on the cool-season whole crop faba legume silage.

Yang et al. [37] and Yang [7] reported that with increasing dosage levels, the NDF digestibility was linearly responsive. The innovative enzyme dosage treatment with 0.5 mL of innovative enzyme per kg of silage DM had the highest NDF digestibility (48.5%) compared with other innovative enzyme treatments. The % milk fat and fat yield were significantly affected by innovative fibrolytic enzyme application. They were linearly differed among the innovative enzyme treatments, being the highest (4.35%, 1.82 kg/d) for low enzyme dosage groups. Compared with the control group (milk yield 41.5 kg/d with %4.35 milk fat), the innovative enzyme treatments linearly affected and tended to affected milk yields in terms of ECM, FCM. The innovative enzyme treatments also linearly impacted the RCM production efficiency (FCM kg per kg of DM intake) and cubically impacted the ECM production efficiency (ECM kg per kg DM intake).

In our studies [7, 37], we demonstrated the positive and beneficial effects of pre-adding and pre-treating the cool-season whole crop faba legume silage with a lower dose level (0.5, 1 mL of FETR per kg DM of cool-season faba silage based TMR) of innovative fibrolytic enzyme to lactation dairy cows which could benefit the development of a new and alternative feeding strategy in western Canada.

5. Recent research in using advanced vibrational (micro)spectroscopy for cool-season legume and legume-cereal silage research at cellular and molecular levels

The silage's nutritional value and digestive behaviour are affected by not only the chemical composition profile but also molecular structure conformation, and biological component matrix. However, the wet-chemical analysis method fails to reveal internal molecular structure and component matrix due to processing and digestion in wet chemical analysis. Advanced synchrotron-radiation and Globar-sourced vibrational (micro)spectroscopy is capable of revealing internal structure features at cellular and molecular levels and simultaneously provides four kinds of important information: not only chemical composition, but also molecular structure, environment, and chemistry, within intact tissue with a highly spatial resolution [8–15]. The detailed principle and methodology of using synchrotron-radiation and Globar-sourced vibrational (micro)spectroscopy have been reported before [8–15].

Recently, our team [2–4] carried out various studies using advanced molecular spectroscopic techniques, either synchrotron-based [38] or global based molecular spectroscopy to reveal (1) the impact of cool season low-tannin (cv. CDC Snowdrop) and high-tannin faba varieties (cv. CDC SSNS-1); (2) the impact of maturity cutting stage at 88-d mid-pod cutting stage and 97-d late-pod cutting stage on inherent structure spectral profile of cool-season whole crop faba legume silage at a molecular level; (3) investigate the interactive association and relationship between molecular structural profiles and nutrient utilisation and availability in ruminant livestock systems [39].

Guevara [4] applied molecular spectroscopic technique by using ATR-FTIR vibrational spectroscopy to study protein and carbohydrate structure make-up for the cool-season faba legume silages and compare different cool-season faba varieties with different tannin levels. It was found that the cool-season low-tannin faba silage had a higher total carbohydrates (TC) spectral peak area at the late-pod cutting stage than at the mid-pod cutting stage with a difference of +3.45 AU. For the structural carbohydrates spectral area intensity (STC), the cool-season low-tannin faba silage was higher when cut at mid-pod cutting stage (difference: +4.11 AU) than the cool-season high-tannin faba legume silage at late-pod cutting stage.

Amides functional group study [4] showed that the low-tannin silage had decreased the amide I area (−1.40 AU) when cut at mid-pod cutting stage than that at the late pod stage. As to amide II structure profile, the cool-season high-tannin faba legume silage had higher amide II at the late pod cutting stage than the high-tannin silage cut at the mid-pod stage with different +2.50 AU.

We also carried out detailed PCA spectral analyses of all the carbohydrate-related spectral region (ca. 879–1485 cm^{-1}). The 83% of the total variation was explained by PC1. In this region, it includes NSTC, TC, and STC regions. The results showed dramatical difference in the cool-season whole crop faba legume silage when cut at mid-pod stage or cut at late-pod stage. It is interesting to find that starch level in cool-season faba legume silage has a strongly positive correlation with structural carbohydrate peak number four (STC4) spectral height intensity with $r = 0.94$.

Protein 2nd structure spectral profile study showed that total digestible nutrients TDN, bio-energy value ($r = 0.76$), and crude protein level ($r = 0.62, 0.65$) in the cool-season faba legume silage positively correlated to α-helix and β-sheet. The TDN and bio-energy values also strongly positively correlated ($r = 0.85$) with the amide I spectral area.

The rumen undegradable protein and rumen bypass starch were strongly correlated to the structural carbohydrate spectral peak number one (STC1) height in the cool-season faba legume silage (RUP; $r = -0.82$; BSt; $r = -0.84$), while, rumen undegradable protein (RUP; $r = -0.83, -0.90$) was strongly negatively correlated with amide I peak height (RUP; $r = -0.83$) and STC area (RUP; $r = -0.90$), as well as α-helix to β-sheet spectral peak height ratio (RUP; $r = -0.73$).

The relationship between molecular structure and intestinal digestions and nutrient supply study showed that intestinal digested crude protein (IADP) and metabolizable protein (MP) in cool-season faba silage were strongly correlated with structural carbohydrates peak # one (STC1) spectral height (IADP; $r = -0.90$; MP; $r = -0.92$). For MP value, it also strongly positively correlated to the protein 2nd structure profile in terms of α-helix peak height, β-sheet peak height, and amide I area ($r = 0.86, 0.86, 0.71$, respectively). The cool-season silage feed milk value based on the DVE value (FMV^{DVE}) and silage feed milk value based on MP (FMV^{NRC}) were strongly correlated to cellulosic compound (CEC) spectral area ($r = -0.95, -0.82$, respectively).

Our results showed that cool-season faba silage starch content could be predicted using α-helix peak heights, amide I, and STC4 with good estimation power ($R^2 > 0.96$), but total digestible nutrients, net energy of lactation, and crude protein were predicted by above molecular structure profiles with no good estimation power ($R^2 < 0.67$). On the other hand, important rumen kinetics, intestinal and total tract digestibility, and metabolic features were highly related to spectral areas of STC, CEC, and amide which can be used to predict with good estimation power ($R^2 > 0.74$).

6. Summary and conclusion

Based on the scientific findings presented in this chapter, the following most important conclusions can be drawn:

1. Cool-season faba (legume) variety with different tannin levels impacts not only the nutrient profile but also protein and carbohydrate-related molecular structure makeup.

2. Additionally, the nutrient supply, bioenergy, degradation, digestion, and metabolic characteristics of cool-season faba silage and intercropped faba-oat silage were highly related independently and synergistically to molecular structure conformation.

3. Furthermore, the nutrient utilisation and availability of cool-season faba silage and intercropped silage in ruminant livestock systems could be accurately predicted by the protein and carbohydrate molecular structures revealed with cutting stage vibrational molecular spectroscopy when they work together.

4. Additive fibrolytic enzyme and maturity stage at harvesting significantly impacted both nutritional and molecular structural changes of legume and legume-cereal intercropped silage.

5. Dairy production performance and milk yield (ECM, FCM, fat yield) studies showed whole plant faba legume silage in early lactating cows could be used as alternative silage to replace traditional barley and corn silages.

Acknowledgements

We want to thank David A. Christensen, Bunyamin Tar'an (CDC, Plant Science), John McKinnon, Bart Lardner, Zhiyuan Niu, and Brent Barlow (CDC) for their technical assistance. This chapter is part of three graduate theses (project abstracts from VG, CN, and JY) and made revisions for book chapter.

Funding

The Chair (PY) feed research programs have been supported by the Ministry of Agriculture Strategic Feed Research Chair Programs, the Natural Sciences and Engineering Research Council of Canada (NSERC, Canadian federal government), the Saskatchewan Pulse Growers (SPG), the Prairie Oat Grower Association, SaskCanola, the Saskatchewan Agricultural Development Fund (ADF), AB Vista (UK), and Various Feed and Animal Industries etc.

Authors' contributions

VG, CN, JY, JH, ME, WZ, TR, and PY wrote, reviewed, edited, and approved the book chapter.

Competing interests

The authors declare that we have no competing interests.

Ethics approval and consent to participate

Not applicable.

Consent for publication

Not applicable.

Availability of data and materials

Not applicable.

Abbreviations

BST	Rumen bypass starch
CDC	Crop Development Centre
CEC	Cellulosic compound
CNCPS	Cornell Net Carbohydrate and Protein System
DVE	total intestinal digestible protein supply with DVE/OEB system
ECM	energy corrected milk
ED_N/ED_CHO	rumen available N and rumen available carbohydrates (ED_N/ED_CHO) hourly effective degradation ratios
FCM	fat corrected milk
FE	fibrolytic enzyme
FMV	feed milk value
IADP	intestinal absorbable protein
ME	metabolizable energy
MP	total metabolizable protein
NE_L	net energy of lactation
OEB	degraded protein balance
PCA	principal component analysis
RUP	rumen undegraded protein
SR-IMS or SR-FTIRM	synchrotron-based infrared microspectroscopy
STC	structural carbohydrate
TDN	total digestible nutrients
TDST	total tract digested starch
VFA	volatile fatty acid

Author details

Victor Guevara[1], Carlene Nagy[1], Jen-Chieh Yang[1], Jiangfeng He[2], Maria E. Rodriguez-Espinosa[1], Weixian Zhang[3], Tao Ran[4] and Peiqiang Yu[1*]

1 Department of Animal and Poultry Science, College of Agriculture and Bioresources, University of Saskatchewan, Saskatoon, Canada

2 Inner Mongolia Academy of Agriculture and Husbandry Science, Hohhot, China

3 Henan University of Animal Husbandry and Economy, Zhengzhou, China

4 College of Pastoral Agriculture Science and Technology, Lanzhou University, Lanzhou, China

*Address all correspondence to: peiqiang.yu@usask.ca

References

[1] Saskatchewan Pulse Growers. Chickpea and faba Bean Situation [Online]. 2018. Available from: https://saskpulse.com/files/report/180727_Low_Res_Penner.pdf [2023 May 05]

[2] Rodríguez-Espinosa ME. Effect of Varieties/Lines, Processing Methods, and Tannin Levels (Low and Normal) on Structural, Physicochemical, and Nutritional Characterization of Faba Bean Grown in Western Canada. MSc. Thesis. Canada: University of Saskatchewan; 2018

[3] Rodríguez-Espinosa ME. Investigation of the Role of Amide I to Amide II Ratio and Alpha Helix to Beta Sheet Ratio of Faba Bean Seeds in the Determination of Microbial Protein Synthesis and Animal Performance and Metabolism in Ruminant Livestock Systems. PhD Thesis. Canada: University of Saskatchewan; 2023

[4] Guevara V. Molecular Structural and Nutritional Evaluation of Faba Bean Plants as Hay and Silage for Ruminants: Effect of Tannin Level, Cutting Stage, and Frost-Damage. PhD Thesis. Saskatoon, Canada: University of Saskatchewan; 2020

[5] Mcallister TA, Hristov AN, Beauchemin KA, Rode LM, Cheng KJ, Bedford MR, et al. Enzymes in ruminant diets. 2000. DOI: 10.1079/9780851993935.0273

[6] McAllister TA, Oosting SJ, Popp JD, Mir Z, Yanke LJ, Hristov AN, et al. Effect of exogenous enzymes on digestibility of barley silage and growth performance of feedlot cattle. Canadian Journal of Animal Science. 1999;**79**:353-360. DOI: 10.4141/A98099

[7] Yang J-C. Effect of Fibrolytic Enzymes on Lactational Performance, Feeding Behavior and Digestibility in Lactating Dairy Cows Fed a Whole Plant Faba Bean Silage-Based Diet. MSc Thesis. Saskatoon, Canada: University of Saskatchewan; 2021

[8] Miller L. Infrared Spectroscopy and Imaging. [Online]. 2002. Available from: http://www.nsls.bnl.gov/newsroom/publications/otherpubs/imaging/workshopmillerhighres.pdf [August 2008]

[9] Marinkovic NS, Huang R, Bromberg P, Sullivan M, Toomey J, Miller LM, et al. Center for Synchrotron Biosciences' U2B beamline: An international resource for biological infrared spectroscopy. Journal of Synchrotron Radiation. 2002;**9**:189-197

[10] Marinkovic NS, Chance MR. Synchrotron infrared microspe-ctroscopy. In: Meyers RA, editor. Encyclopedia of Molecular Cell Biology and |Molecular Medicine. Weinheim: Wiley-VCH Verlag GmbH & Co. KGaA; 2006. pp. 671-700

[11] Doiron KJ, Yu P, McKinnon JJ, Christensen DA. Heat-induced protein structures and protein subfractions in relation to protein degradation kinetics and intestinal availability in dairy cattle. Journal of Dairy Science. 2009a;**92**:3319-3330

[12] Doiron KJ, Yu P, Christensen CR, Christensen DA, McKinnon JJ. Detecting molecular changes in Vimy flaxseed protein structure using synchrotron FTIRM and DRIFT spectroscopic techniques: Structural and biochemical characterization. Spectroscopy. 2009b;**23**:307-322

[13] Wetzel DL, Eilert AJ, Pietrzak LN, Miller SS, Sweat JA. Ultraspatially-resolved synchrotron infrared

microspectroscopy of plant tissue *in situ*. Cellular and Molecular Biology. 1998;**44**(1):145-167

[14] Wetzel DL, LeVine SM. Biological applications of infrared microspectroscopy. In: Gremlich H-U, Yan B, editors. Infrared and Raman Spectroscopy of Biological Materials. Practical Spectroscopy Series. Vol. 24. NY: Marcel Dekker; 2001. pp. 101-142

[15] Wetzel DL, Srivarin P, Finney JR. Revealing protein infrared spectral detail in a heterogeneous matrix dominated by starch. Vib. Spect. 2003;**31**:109-114

[16] Nagy C. Physiochemical and Nutritional Evaluation of Intercropping Whole Plant Faba Bean with Whole Plant Oat as Hay and Silage for Ruminants: Effects of Growth Stage/Cutting Time and Adding Fibrolytic Enzyme from Trichoderma Reesei. MSc Thesis. Saskatoon, Canada: University of Saskatchewan; 2023

[17] NRC. Nutrient requirement of dairy cattle. 7th ed. Washington, DC: National Research Council, National Academy of Science; 2001

[18] National Research Council (NRC). Nutrient Requirement of Beef Cattle. 7th ed. Washington, DC: NRC, National Academy of Science; 2016

[19] Higgs RJ, Chase LE, Ross DA, Van Amburgh ME. Updating the Cornell net carbohydrate and protein system feed library and analyzing model sensitivity to feed inputs. Journal of Dairy Science. 2015;**98**:6340-6360. DOI: 10.3168/jds.2015-9379

[20] Van Amburgh ME, Collao-Saenz EA, Higgs J, Ross DA, Recktenwald EB, Raffrenato E, et al. The Cornell net carbohydrate and protein system: Updates to the model and evaluation of version 6.5. Journal of Dairy Science. 2015;**98**:6361-6380. DOI: 10.3168/jds.2015-9378

[21] Ørskov ER, McDonald I. The estimation of protein degradability in the rumen from incubation measurements weighted according to rate of passage. Journal of Agriculture Science (Cambridge). 1979;**92**:499-503

[22] Tamminga S, Van Straalen WM, Subnel APJ, Meijer RGM, Steg A, Wever CJG, et al. The Dutch protein evaluation system: The DVE/OEB-system. Livestock Production Science. 1994;**40**:139-155

[23] Tamminga S, Brandsma GG, van Duinkerken GG, van Vuuren AM, Blok MC. Protein Evaluation for Ruminants: The DVE/OEB 2007-System. CVB Documentation Report. Wageningen, NL: Wageningen University; 2007. pp. 53-58

[24] Tas BM, Taweel HZ, Smit HJ, Elgersma A, Dijkstra J, Tamminga S. Effects of perennial ryegrass cultivars on milk yield and nitrogen utilization in grazing dairy cows. Journal of Dairy Science. 2006;**89**:3494-3500. The Canadian Encyclopedia [Online] Available from: http://www.thecanadianencyclopedia.ca/en/article/faba-bean/ [2021 Oct. 23]

[25] Tamminga S, van Vuuren AM, van der Koelen CJ, Ketelaar RS, van der Togt P. Ruminal behaviour of structural carbohydrates, non-structural carbohydrates and crude protein from concentrate ingredients in dairy cows. Netherlands Journal of Agricultural Science. 1990;**38**:513-526

[26] Gargallo S, Calsamiglia S, Ferret A. Technical note: A modified three-step in vitro procedure to determine intestinal digestion of proteins. Journal of Animal Science. 2006;**84**(8):2163-2167. DOI: 10.2527/jas.2004-704

[27] Guevara-Oquendo VH, Christensen DA, Refat B, Rodriguez Espinosa ME, Feng X, Yu P. Production performance and metabolic characteristics of cows fed whole plant Faba bean silage in comparison with barley and corn silage. Canadian Journal of Animal Science. 2022;**102**:145-154. DOI: 10.1139/CJAS-2021-0048

[28] Association of Official Analytical Chemists (AOAC). Official Methods of Analysis. 18th ed. Gathersburg, MD, USA: AOAC; 2005

[29] Refat B. Molecular Structure Features and Nutrient Availability and Utilization of Barley Silage Varieties with Varying Digestible Structural Carbohydrate in Comparison with A New Short-Season Corn Silage in High-Producing Dairy Cattle. PhD. Thesis. Saskatoon, Canada: University of Saskatchewan; 2018

[30] Refat B, Christensen DA, McKinnon JJ, Yang W. Effect of fibrolytic enzymes on lactational performance, feeding behavior, and digestibility in high-producing dairy cows fed a barley silage-based diet. Journal of Dairy Science. 2018;**101**(9):7971-7979. DOI: 10.3168/jds.2017-14203

[31] Yang WZ, Beauchemin KA, Rode LM. Effects of an enzyme feed additive on extent of digestion and Milk production of lactating dairy cows. Journal of Dairy Science. 1999;**82**:391-403

[32] Bowman GR, Beauchemin KA, Shelford JA. The proportion of the diet to which fibrolytic enzymes are added affects nutrient digestion by lactating dairy cows. Journal of Dairy Science. 2002;**85**:3420-3429

[33] Peters A, Meyer U, Dänicke S. Effect of exogenous fibrolytic enzymes on performance and blood profile in early and mid-lactation Holstein cows. Animal Nutrition. 2015;**1**:229-238

[34] Peters A, Lebzien P, Meyer U, Borchert U, Bulang M, Flachowsky G. Effect of exogenous fibrolytic enzymes on ruminal fermentation and nutrient digestion in dairy cows effect of exogenous fibrolytic enzymes on ruminal fermentation and nutrient digestion in dairy cows. Archives of Animal Nutrition. 2010;**64**:221-237

[35] Iiyama K, Lam TBT, Stone BA, B.A. Covalent cross-links in the Cell Wall. Plant Physiology. 1994;**104**:315-320

[36] Yang J-C, Guevara-Oquendo VH, Refat B, Yu P. Effects of exogenous Fibrolytic enzyme derived from Trichoderma Reesei on rumen degradation characteristics and degradability of low tannin whole plant Faba bean silage in dairy cows. Dairy. 2022a;**3**:303-313. DOI: 10.3390/dairy3020023

[37] Yang J-C, Guevara-Oquendo VH, Refat B, Yu P. Lactational performance, feeding behavior, ruminal fermentation and nutrient digestibility in dairy cows fed whole plant Faba bean silage-based diet with Fibrolytic enzyme. Animal: The International Journal of Animal Biosciences. 2022b;**16**: In Press. DOI: 10.1016/j.animal.2022.100606

[38] Miller LM, Dumas P. Chemical imaging of biological tissue with synchrotron infrared light. Biochimica et Biophysica Acta. 2006;**1758**:846-857

[39] Jonker A. Characterization of Anthocyanidin-Accumulating Lc-Alfalfa for Ruminants: Nutritional Profiles, Digestibility, Availability and Molecular Structure, and Bloat Characteristics. PhD Thesis. Saskatoon, Canada: University of Saskatchewan; 2011

Chapter 9

The Utilization of Prairie-Based Blend Pellet Products Combined with Newly Commercial Phytochemicals (Feed Additives) to Mitigate Ruminant Methane Emission and Improve Animal Performance

Taufiq Hidayat, Maria Eugenia Rodriguez Espinosa, Xiaogang Yan, Katerina Theodoridou, Samadi, Quanhui Peng, Bin Feng, Weixian Zhang, Jiangfeng He and Peiqiang Yu

Abstract

The objective of this review is to comprehensively upbring the development potency of value-added pellet products from prairie industry by-products or co-products in combination with newly developed hydrolysable tannins (HT) and saponin to mitigate ruminant methane emission and improve the productivity of ruminant animals. The prairie region often produced plentiful amount of co-products and by-products that still have nutritional properties and can be utilized as ruminant feed to keep the sustainability in the agriculture sector. In ruminants, rumen microbial fermentation produces methane (CH_4) as one of the outputs that can cause energy loss and act as a potent greenhouse gas (GHG) in the open atmosphere. Recently, the newly developed HT extracted from nutgall *(Gallae chinensis)* and saponin extracted from tea *(Camellia sinensis)* products are commercially available at affordable prices and are able to reduce methane emissions. Reducing methane emissions is vital to aid and support carbon reduction goals, but it must be accomplished while preserving and increasing business, maximizing profit, and providing economic return and benefit to pulse, cereal, and oil-crop growers. In conclusion, the prairie unused product combined with the aforementioned phytochemicals can be developed as a new pellet product. However, further research may be needed to determine the most effective additive levels of both saponin and HT products due to their anti-nutritional abilities while maintaining and improving livestock productivity.

IntechOpen

Keywords: blended pellet product, feed additives, hydrolysable tannin, Saponin, methane mitigation, animal performance

1. Introduction

1.1 Canadian prairies pothole region (CPPR)

Saskatchewan (SK), Alberta (AB), and Manitoba (MB) are three Canadian Prairies with expansive areas, partially covered by grasslands, plains, and lowlands stretching from Alberta's Rocky Mountain foothills to Manitoba's Red River Valley. The Canadian section of the prairie region is the country's largest and most intensive grain crop production area, which spans 312,746 km^2 and accounts for about 83% of Canada's total agricultural land area and about 5% of Canada's total land [1–3]. Based on The Canada Guide [4], the economy of prairie land in Canada is significantly increasing led by the growth of industries followed by jobs and population in the mid-twentieth century with its main industries of services, oil, agriculture [livestock industries (dairy cattle, beef cattle, and sheep), and crop cultivation (canola, wheat, oats, and barley)]. A Recent report released by Canada Agriculture Census [5] also showed that Canadian Prairies (Manitoba, Alberta, and Saskatchewan) accounted for 83% of total Canada's farms, nearly all of Canada's canola (99,2%), spring wheat (97,6%), barley (96,2%), and 72,7% of Canada's cattle industry.

1.2 Canadian prairies feed source potential

Being one of the world's top producers and exporters of agricultural goods, Canada has the potential to play a significant role in the development of the cellulosic biorefinery industry supported by its abundant supply of cellulosic biomass produced by the agricultural sector [6–8]. Canadian Prairie Region produces various types of unused goods that can still be utilized as animal feed due to its high energy and starch contents [9]. The Pulse industry often produces low grade peas (*Pisum sativum*) or lentil screenings as by-products. Peas are processed in various ways such as frozen, raw, or canned, and expelled without using the pod (peas exterior component) and make up around 35–40% of the total peas' weight [10]. With 254.2 g/kg CP, 869.0 g/kg DM, 31.4 g/kg ash, 8.5 g/kg ether extract, and 12.8 MJ/kg EM making peas (*Pisum sativum*) one of the most valuable feed sources either for ruminants or poultry [11]. In bio-oil processing, canola or carinata meals are also produced as co-products. Canola meals are considered as a proper ruminant feed because it is highly palatable to ruminants, inexpensive, it has a well-balanced amino acid (AA) profile and has no direct food value for humans [12–14]. Its protein content has also been proven to be highly degradable in the rumen [15], making it less effective as a post-ruminal AA source (44.3–74%) [16]. However, it is not recommended for livestock to directly consume it without pre-processing due to its poor quality, lack of phytonutrients, and incomplete nutrient content. Nevertheless, supplementing it with a multi-nutrient additive can produce high-quality feed that is able to successfully satisfy the livestock's daily need for nutrients [17]. Additionally, plant-based meal (e.g., soy protein, pea protein, and starches) utilization for animal feed is environmentally friendly because every kilogram of their production releases approximately 1 kg of carbon dioxide into the atmosphere [18].

1.3 Benefit of prairies co-product and by-product utilization

Development of international and domestic markets for prairie pulse, cereal, and oil-crop producers, and feeds and livestock industries is a key to maintain and increase business, maximize profit, and provide economic return and benefit to pulse, cereal, and oil-crop producers. The utilization of both agricultural co-product and by-product based on prairie co-products from bio-oil processing (canola or carinata meal), pulse screenings (damaged peas/lentil, a non-food grade of peas/lentil/faba) can keep the sustainability in the agriculture sector. Moreover, the environmental impact of feed and animal production, as well as the economic value of innovative feeds in alternate applications, is critical [19]. The viability of utilizing alternative feeds for grazing animals is determined by factors such as feed value of novel feeds, animal production responses, and feed costs in comparison with standard diets. Many studies have proven that improper utilization of agricultural waste can cause severe environmental issues such as groundwater pollution, pathogen proliferation, and greenhouse gas emission [20–22]. To improve the competitive market (both domestic and international), it is necessary to establish a new suitable product that is environment-friendly and capable of reducing Greenhouse Gas (GHG) emission by mitigating ruminants (dairy, beef cattle, or sheep) methane, but also these new products have high feed milk/meat value (FMV) and are easily transported/shipped. High production ruminants (dairy and beef cattle) need to have an optimized nutrient supply for optimized high milk/meat production from newly developed feed products without causing severe GHG pollution by mitigating ruminant (dairy, beef cattle, or sheep) methane.

2. Feed additive utilization in ruminant daily feed

2.1 Feed additives on nutrition and performance

Feed additives are chemicals, microorganisms, or preparations that are purposely added to feed or water in order to carry out one or more of the activities mentioned above, besides feed material and premixtures. There are various functions of feed additives including to positively influence the properties of feed, the properties of animal products, and color, fulfil the nutritional needs of animals, influence the environmental effects of animal production, influence animal performance or welfare, particularly by influencing the microorganisms in the gastrointestinal tract or the digestibility of feed, or have a coccidiostatic or histomonostatic effect. Generally, additive substances may be categorized as technological (e.g., preservatives, antioxidants, emulsifiers, stabilizing agents, acidity regulators, silage (grass or other green fodder compacted and stored in airtight conditions, typically in a silo) additives); sensory (e.g., flavorings, colorants); nutritional (e.g., vitamins, minerals, amino acids, trace elements); zootechnical (e.g., digestibility enhancers, gut flora stabilizers); coccidiostats; and histomonostats [23]. Based on its function, there are two types of feed additives: nutritive feed additive and non-nutritive feed additive. Nutritive feed additives are compounds added to the feed ration to improve the nutrient values (e.g., amino acid, vitamin, and mineral), while non-nutritive feed additives are compounds added to the ration to improve values other than nutrients such as palatability (by adding color and odor), preserve the feed quality (by adding antioxidant), or as a pathogen inhibitor [23].

Adding antibiotic feed additives to ruminant diets during the reproductive period can improve absorption of nutrients and reproductive performance, which also has resulted in various positive health advantages. However, their use in ruminant diets is debatable due to the possibility of their deposition into meat and milk as well as the expansion of antimicrobial resistance brought on by the misuse of antibiotics, which has drawn attention to the need for new antibiotic alternatives in the field of animal nutrition [24–26]. Even in countries such as Indonesia and EU, feed additives that contain antibiotic growth promotors (AGP) are banned. This has drawn the researcher's attentions to find alternative feed additives from natural sources, such as herbs and spices, which are affordable, effective, and eco-friendly. Many studies of feed additives have been established and it is proven that adding feed additives to the diet can increase performances in ruminant animals. The summary of several additives piloted to ruminant animals is outlined in **Table 1**.

2.2 Tannin utilization and benefit as ruminant feed additive

Tannins, usually called tannic acid, are a group of phenolic compounds that are regularly found in woody flowering plants used to deter herbivores from consuming them. Tannins have both positive and negative impacts when applied. There are various positive impacts including enhanced protein consumption, rapid body weight gain or wool production, higher milk production, increased fertility, and improved animal well-being and comfort through the reduction of worm loads and the prevention of bloat [47]. According to Goel et al. [48], tannins may be toxic to certain rumen microbes and may have negative effects on ruminant metabolism [49]. Low palatability and impaired diet digestibility are further negative consequences that have been linked to decreased performance [50, 51]. However, the source and concentration of tannins are the main factors that determine whether they are beneficial or not [52].

Tannins are varied among plants and primarily differentiated based on their molecular structure including hydrolysable tannins (HT; polyesters of gallic acid and different individual sugars), condensed tannins (CT; polymers of flavonoids), and mixtures of these two fundamental structures [53]. Condensed tannins are oligomeric and polymeric proanthocyanidins formed by polymerization of flavan-3-ols. CT cannot easily separate because it possesses protein-binding ability, which are flavonoid units linked by carbon-to-carbon bonds and cannot be separated by hydrolysis [54]. The main components of HT are gallotannins and ellagitannins, which can be easily separated by acids, bases, and enzymes [55]. When consumed, HT do not show anti-nutritional effects and give health benefits to livestock feed because of their strong antibacterial, anti-inflammatory, antioxidant, and anti-parasitic effects in animals [56].

2.3 Saponin utilization and benefit as ruminant feed additive

Saponins are secondary compounds that have extensively abundant supply in nature and usually known as non-volatile, surface-active compounds. The word "saponin" is derived from the Latin word "sapo," which means "soap." This is because when saponin molecules are combined with water, foam is formed. Saponins have been found in over 100 plant families, in several marine sources, and even there are a small number of fish that produce saponins as shark repellants [57]. The primary role of saponin is providing defense against many pathogens and herbivores [58–60]. Saponins are usually located in tissues that are most susceptible to bacterial or fungal infection or insect predation. There are three main categories of saponin:

Feed additive type(s)	**Experiment object**	**Effects**	**References**
Green tea extract (GTE)	Buffalo bulls	Fertility rate improves by 16.56%	Ahmed et al. [27]
Yeast	Dairy cattle	Health and productivity improvement	Miller-Webster et al. [28]
Phytochemicals, nitrate (NO_3^-)	Beef cattle	Reduce CH_4 production	Alemu et al. [29]
Apple bagasse yeast	Non-lactating rumen fluid of dairy cattle	Increases feed consumption	Castillo-Castillo et al. [30]
Yeast	Lactating dairy cattle	Milk production, DMI, and live weight gain	Dann et al. [31]
Yeast	Lactating ruminal fluid of dairy cows (Jersey cows)	Improving DMI, NDF digestibility, and naturally modifying rumen fermentation	Lila et al. [32]
Fumaric acid	Beef cattle	Increases total VFA production	Beauchemin and McGinn [33]
Sunflower oil	Angus heifer	Increases digestibility energy intake and the rate of gain cattle, reduces CH_4	Beauchemin et al. [34]
Acacia mearnsii (black wattle) tannin	Sheep	Decreases ruminal ammonia, urinal nitrogen, and methane production	Carulla et al. [35]
Lactobacillus spp.	Pre-ruminant calves	Decreases coliform count, reduces scouring, improves feed intake, liveweight gain	Beeman [36]; Gilliland et al. [37]; Lee and Botts [38]
Lactobacillus spp.	Pre-ruminant lambs	Lower mortality, improves feed intake and liveweight gain	Pond and Goode [39]; Umberger et al. [40]
Garlic and citrus extract	Sheep	Decreases methane emission, increasing ruminal activity	Ahmed et al., [41]
Hemicellulose extract	Dairy cattle	Improve fiber degradation	Herrick et al. [42]
Microalgae and rapeseed meal	Dairy cattle	Improve fiber and nitrogen digestibility, increase NH_3 concentration	Lamminen et al. [43]
Tannin	Beef cattle	Reduces NH_3 and CH_4, improves propionate and butyrate concentration	Orzuna-Orzuna et al. [44]
Lipid	Beef cattle	Increases propionate molar proportion, lowering acetate molar proportion and VFA	Dai and Faciola [45]
Macroalgae (*Sargassum fulvellum*)	Beef cattle	Increase DMD, total gas emission, and VFA	Choi et al. [46]

DMI: dry matter intake, NDF: neutral detergent fiber, VFA: volatile fatty acid, DMD: dry matter digestibility.

Table 1.
Various feed additive sources and their effects on ruminants.

triterpenoid saponin, steroid saponin, and alkaloid saponin. Triterpenoid saponin is the most distributed in the plant kingdom and it is a phrase that denotes three monoterpene molecules, each of which has three carbon atoms. This indicates that there are six molecules totaling 30 carbon atoms [61]. Triterpene saponin consists of two types (e.g., monodesmosidic and didesmosidic), where mono- and didesmosidic have single and double sugar chain, respectively [62, 63]. Steroid saponin is a type of triterpenoid saponin that has undergone modification. Its structure is made up of 27 carbon atoms in bicyclic five-membered rings and tetracyclic six-membered rings. Alkaloid saponin have structure similar to steroid saponins, the only difference is that the alkaloid saponin has piperidine ring (a six-membered ring carrying N atom) rather than pyranose ring (a six-membered ring carrying O atom) [61]. Numerous activities of saponin (e.g., antimicrobial, antihelminthic, insecticidal, larvicidal, and molluscicidal) have already been documented [64]. In ruminant animals, dietary saponins have significant effects on all phases of metabolism, including feed ingestion and waste excretion [65]. Also, it has been reported that saponins are effective antifungal and antiviral agents [66]. Several sources of saponins have been discovered to be devastating to protozoa and have been named as potential defaunating agents in the rumen [67, 68].

2.4 Pellet processing effects on value-added product

Pelleting is the process of forcing a pulverized mixture of feed materials through a metal plate with cylindrical holes [69]. Pelleting is one of the ways to reduce particle size to accelerate nutrient fermentation in rumen. Gustafson [70] characterized the forces occurring on the pellets as impact, compression, and shear; impact forces break the pellet outer layer and any existing cleavage planes in the pellet; compression forces crush the pellet and create failure along cleavage planes; shear pressures abrade the pellet's corners and exterior. Reducing feed particle size in daily livestock feed has different impacts on digestion that affect each other: (1) increasing dry matter intake (DMI), (2) increasing the surface area for bacteria to attach, resulting in improvement of ruminal degradation, (3) affecting chewing time and saliva production, which have further effects on ruminal pH because saliva acts as a buffer, and (4) affecting rumen retention, which possibly supports the improvement of subacute ruminal acidosis (SARA) [71–73]. Furthermore, blended feed substances processed into pellets can balance amino acid delivery, enhance and optimize nutritional supply, and alter rumen fermentation behavior [74]. Pelleting has various technical advantages, including enhanced stability (because of very low moisture content) and simpler handling, storage, and transportation [71]. Additionally, Johnson and Johnson [75] reported that grinding or pelleting forage diets have been shown to lower enteric CH4 emissions by 20–40% at high intakes. This might be explained by the faster rate of feed transit, which reduces the amount of time the feed is exposed to ruminal digestion [76].

3. Ruminant methane emission

3.1 Methane emission mechanism

Methane also known as marsh gas or methyl hydrate was discovered and isolated by Alessandro Volta in November 1776 in Lake Maggiore, Italy. Methane is a colorless and fragrantless gas widely found in nature as a result from the decay/decomposition of organic matter by certain bacteria and usually used by humans as fuel to make heat and light. Methane is the most basic of the paraffin series of hydrocarbons and the simplest member of the alkane family, which is a group of organic compounds consisting only of carbon and hydrogen atoms and one of the most potent greenhouse gases (GHG), and it has the molecular formula CH_4. According to Britannica [77], the characteristic of methane has a specific gravity of 0.554, making it lighter than air, hardly dissolves in water but dispersible in organic solvent, and quickly burns in the presence of air; releasing carbon dioxide and water vapor, the flame is fierce, pale, and barely bright, has a melting point of 182.5°C (296.5°F) and a boiling temperature of 162°C (259.6°F). The general methane formation equation is:

$$CO_2 + 8\mathrm{H}^+ + 8\mathrm{e}- \rightarrow CH_4 + 2\mathrm{H_2O} \tag{1}$$

Methane can also be produced by polygastric animal such as cows and lambs as a natural by-product of the digestion and fermentation that occurs in the ruminal guts (rumen) through a process called methanogenesis. Methanogenesis is an anaerobic reprocess where C atom contained in CO or CO_2 reduced to CH_4, with the intention to avoid hydrogen accretion, which subsequently inhibits dehydrogenase enzyme activity and disturbs the fermentation mechanism [78]. According to Janssen and Kirs [79], there are 113 species and up to 28 genera of methanogens that have been discovered in nature, while *Methanobrevibacter* (61,6%) is regarded as the most dominant methanogen in the rumen. Methanogens are the primary component of the *Euryarchaeota* and are separated into five orders including *Methanococcales*, *Methanobacteriales*, and *Methanomicrobiales* [80]. These methanogenic bacteria already exist in cows, even in the stage of pre-ruminant [81, 82]. Methane is mostly produced in rumen (80–95%), while other small quantity is produced in large intestine (5–20%). Methanogenesis can occur *via* CO_2 reduction utilizing H_2 as an electron source, methyl-group reduction, or acetate reduction [83].

Ruminant animals consume plant materials as their primary source of nutrition that contains structural carbohydrates, proteins, and other feed components (**Figure 1**). These complex structures are hydrolyzed to simpler monomers, and then are subsequently fermented by rumen microorganisms to produce VFA (acetate, propionate, butyrate, and small amount of valerate), CO_2, CH_4, and H_2 [85]. Under anaerobic conditions in the rumen, oxidation reactions require ATP to release hydrogen. The amount of hydrogen produced is highly dependent on the type of feed and the type of microbes that work to ferment the feed in the rumen [78]. The methanogenic archaea bacteria and other microorganisms that reside in the rumen utilize hydrogen (H_2) that has been mainly produced during hydrogenase microbial fermentation, carbon dioxide (CO_2), and a certain amount of intermediate fermentation products that have been produced by other microbes as substrates to generate methane (CH_4), which is their only method for energy acquisition.

There are three major pathways for rumen fermentation: the hydrogenotrophic pathway converts H_2 and CO_2 produced by bacteria, fungi, and protozoa into CH_4 [86, 87]. The most prevalent hydrogenotrophic bacteria are from the genus Methanobrevibacter, which is classified into two groupings, the SGMT clade (Mbb. gottschalkii, Mbb. smithii, Mbb. Thaueri, and Mbb. millerae) and the RO clade (Mbb. ruminantium and Mbb olleyae) [79, 88]; and methyl groups, which are found in methylamines and methanol [89, 90]. Methylamines are formed from glycine betaine

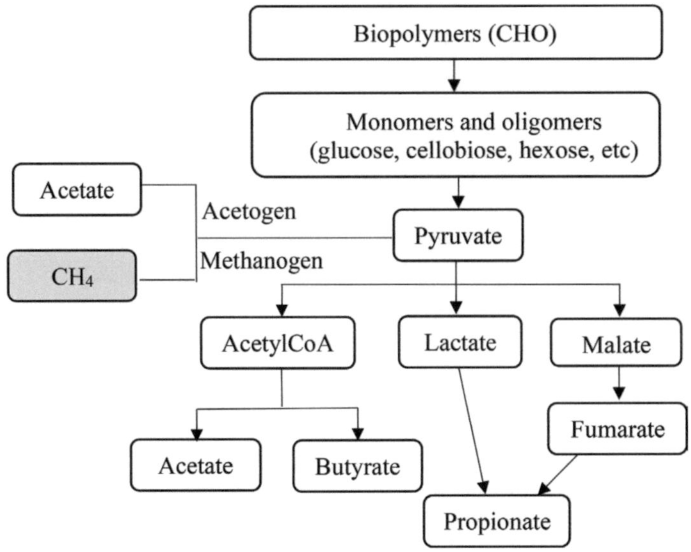

Figure 1.
Methane production in rumen (adapted from [84]).

(derived from beetroot) and choline (which is found in plant membranes), whereas methanol is generated from the hydrolysis of methanolic side groups in plant polysaccharides; the aceticlastic pathway is reviewed by Morgavi et al. [91]. As for methane excretion, ruminants have unique digestive system that allows them to regurgitate and re-chew their partially fermented feed as cud. During this process, the accumulated gases including methane are released in the form of exhaust gas (farts and burps) as well as in feces [78, 92].

3.2 Factors affecting ruminant methane emission

3.2.1 Type and quality of feed

Type and quality of feed can influence the synthesis of methane in the rumen. Broucek [93] reported that forage species, forage processing, forage fraction in the diet, and grain supply affect the CH_4 generation in ruminants. These are mostly linked with carbon supply that will affect the whole activities of ruminant microorganisms. Improved feed quality is also intended to improve animal performance. Thus, improved diet quality can be an efficient way of lowering emissions per unit of animal product [76]. Certain feed components, such as high-fiber forages, promote more extensive fermentation and higher methane emissions compared to low-fiber or grain-based diets because fiber-rich feeds require extensive microbial activity to digest, leading to increased methane production. Methane production tends to decrease as feed protein concentration increases, but it also increases when feed fiber content increases [75, 94]. When compared to a lower concentrate diet (around 30 or 40%), a higher concentrate diet (around 80 to 90%) can minimize gross energy loss caused by methane by 2 to 3% [95]. Recent research conducted by Olijhoek et al. [96] on Holstein and Jersey cows fed with high concentrate diet (up to 91%) showed that there was a noticeable connection between breed and diet between Holstein cows and Jersey cows (48 and 22%, respectively). Although dietary adjustments to consume less forage may lower methane generation, they may cause other physiological problems that could possibly devaluate pH and lead to severe ruminal acidosis [97].

3.2.2 Level and feed intake

Enteric methane emissions are clearly linked to dry matter intake (DMI) either in dairy or in beef cattle [98]. According to Shibata and Terada [94], CH_4 generation normally increases as the daily feed intake increases. Generally, when ruminants consume more feed, and their rumen becomes more active, leading to increased fermentation and higher methane production. In the rumen, which is the first chamber of a ruminant's multi-compartment stomach, microbes break down the carbohydrates present in the feed into monomer or oligomer compounds, producing volatile fatty acids (VFAs), which then also produce metabolic by-products. Energy is produced by the ruminants' digestion of carbohydrates, which also produces enteric methane (CH_4) emissions [99]. Methane emissions were lower on a high concentration diet (920 g/kg DM) than on a mixed (forage/concentrate) diet (500 g/kg DM) [100]. When it comes to neutral detergent fiber (NDF) intake, the CH_4 production is higher when cattle consume high-fiber digestibility diets, which can boost the acetic acid (CH_3COOH) production that exceeds the propionic acid (CH_3CH_2COOH). The acetic acid subsequently results in the release of H, which is utilized by methanogens to form CH_4 [76].

3.2.3 Rumen microbial content

The rumen is the primary generator of methane and specific microbiome characteristics are linked to low/high methane levels. In the rumen, ciliate protozoa synthesize H_2, which is the principal substrate for methanogenesis in the rumen, and removing them (defaunation) resulted in 11% less methane emissions [101]. Intuitively, the methanogenic bacteria population should be linked to methane emission. However, some researches in dairy cattle, sheep, and beef cattle showed that there is a weak, or even no correlation between methanogens' overall abundance with methane emission [102–108].

Rumen microbial contents are more likely affected and closely related to the type and composition of feed given to the ruminants. Thus, the feed-contained-nutrient decides the amount of H_2 produced in the process of forming acetate and butyrate and the use of H_2 which can be oxidized to H_2O, accompanied by the reduction of CO_2 to CH_4. Research conducted using steers showed that rumen fluid from concentrate-fed steers had more propionic acid and less acetic acid, as well as less archaea and protozoa than mixed-fed steers; these rumen contents, particularly protozoa and archaea, show a strong association with CH_4 emissions (g/kg DMI) [100, 107].

3.2.4 Environment temperature

In tropical climate regions, which typically have higher temperatures, it affects the quality of forages. The forage's cell wall, acid detergent fiber, and lignin tend to rise, resulting in declined digestibility of feed and increased energy loss, which continuously leads to decreased feed intake and an increase in CH_4 generation due to a drop in animal production efficiency [94, 109, 110]. It is also attributed to extended preservation time in the rumen and a reduced rate of methanogen outflow from the rumen to the abomasum [109]. Furthermore, Lee et al. [110] found that elevated temperatures may result in an increase in methane generation of 0.9% every 1°C of temperature rise and 4.5% every 5°C of temperature rise. Methane generation per DMI rose and was nearly 10% greater at temperatures over 26°C than at 18°C temperatures in cows at the preservation level of feeding [109].

Overheat temperature can cause severe heat stress and also inflict on cattle itself. Cattle usually will drink more and eat less when the temperature rises. Heat stress can increase rectal temperature, respiratory rate, pulse rate, and water intake, while subsequently reduce body weight gain, dry matter intake, and CH_4 emission. A heat stress experiment on ruminants conducted by Yadav et al. [111] using non-lactating crossbreed dairy cows showed that CH_4 emissions fell significantly with increasing temperature up to >35°C. Furthermore, as animals begin to suffer from heat exhaustion, their food intake decreases, and their metabolism slows [112].

4. Ruminant methane mitigation

4.1 Effect of methane production for ruminants

Ruminant livestock production plays a significant role in global agricultural systems, providing a valuable source of meat, milk, and fabrics for humans. However, ruminants, such as cattle, sheep, and goats, are known to produce and release substantial amounts of anthropogenic methane (CH_4) during their digestive process (around 250 to 500 L per day) to the open world and can cause greenhouse effect to the environment that may occur in the next 50–100 years, resulting on climate change and risen the average temperature of the earth [75, 113]. Methane is one of the potent

Ruminant types	CH_4 production
Dairy cows (avg.)	151–497 g/day
• Holstein	299 g/day
• Crossbread	264 g/day
Lactating cows	354 g/day
Non-lactating cows	269 g/day
Heifers (avg.)	223 g/day
• Heifers grazing on fertilized pasture	223 g/day
• Heifers grazing on unfertilized pasture	179 g/day
Dairy ewes	23 g/day
Beef cattle (avg.)	161–323 g/day
• Mature beef	240–396 g/day
• Cattle feed with pasture	230 g/day
• Cattle feed with high grain	70 g/day
Suffolk sheep	22–25 g/day
Bison	200 g/year

Source: ([93]; [122]).

Table 2.
Ruminant CH_4 production.

greenhouse gases produced during the anaerobic fermentation of feed in ruminants, contributing to global warming and climate change of approximately 15% of the world's total methane emission [114]. Methane only lasts for a relatively brief time in the atmosphere (around 8, 4 to 12 years) compared to other greenhouse gases which lasted for a longer period [CO_2 (300–100 years), CFC (40–150 years), N_2O (114 years), SF_6 (3200 years), NF_3 (740 years), HFC (270 years), PFC (2600–5000 years)]. It is reported that methane is 21 times as potent as carbon dioxide (CO_2) at trapping heat in the atmosphere [115–117]. Additionally, methane formed by the ruminants can also inflict around 2 to 12% of energy loss [118], causing feed inefficiency and financial waste. Animal species, DMI, type of forage fed, overall ratio of concentrate to forage, feed conversion efficiency, addition of lipids or ionophores to the diet, plant secondary metabolites, alteration in the ruminal microflora, and rumen fermentation features, such as VFA and hydrogen (H_2), all affect the CH_4 synthesis [75, 119–121]. Furthermore, according to Broucek [93], not only diet but also different types of ruminants can produce different amount of methane emission (**Table 2**).

Strategy	Mechanism	Effects on CH4	Problem
Ionophores	Inhibiting H_2 producer activity	Medium	1. Bacterial resistance 2. Residue
Halogenated compounds	Inhibiting methanogens activity	High	1. Toxic 2. Residues 3. Bacterial resistance
Phytochemicals	A broad antimicrobial activity	Medium	1. Expensive 2. Bacterial resistance 3. Performance decline
Lipids	Inhibiting methanogens activity	High	1. Expensive 2. Negative effects on performance
Nitrooxy compounds	Inhibiting methanogens activity	High	1. Expensive 2. Potential bacterial resistance
Algae	Inhibiting methanogens activity	High	1. Affect rumen fermentation 2. Residue
Propionate precursors	Competing with methanogenesis for hydrogen source	Low	1. Expensive 2. Inefficiency
Concentrates	Competing with methanogenesis for hydrogen source	Medium	1. Costs 2. Acidosis risk
Forages	Lowering CH_4 emissions per unit of meat and milk	—	Increasing the absolute emission
Non-forage fiber sources	Competing with methanogenesis for hydrogen source	Low	Inefficiency

Source: [130, 131].

Table 3.
Methane reduction strategies through diet manipulation.

4.2 Diet manipulation to mitigate ruminant methane

Methane is produced as part of an inevitable and natural rumen fermentation outcome. Over the decades, scientists and researchers have tried numerous methods to suppress the ruminant methane emission, such as production intensification, altering diet management, diet manipulation, rumen manipulation, and selection of low-CH_4-producing animals [123]. Adding feed additives to dietary feed is one of the most common methods conducted by many researchers. Dietary manipulation method can decrease CH_4 emission by 40% [124]. Even in another study, it was found that improved nutrition may allow for a reduction in CH_4 emissions of up to 75% [125]. There are two broad groups of dietary tactics: (1) enhancing the forage quality and adjusting the diet's percentages and (2) feeding chemicals to animals that either directly prevent methanogens or modify metabolic pathways to reduce the substrate for methanogenesis as a feed additive [126]. Notably, there are at least eight dietary intervention types that have been conducted from 2000 to 2020 (i.e., oils, macroalgae, nitrate, ionophores, protozoa controls, phytochemicals, essential oils, and 3-nitrooxypropanols). The development of feed additive made from oregano and green tea extract can reduce CH_4 gas emission in dairy cows [127], feed additive made from the mixture of xylanase and *Saccharomyces cerevisiae* has been proven to lower the CH_4 of agricultural calf farms [128], and feed additive developed from algae; *Ulva sp.* decreased CH_4, NH_3, and VFA production, while *Sargassum horneri* decreased rumen CH_4 and NH_3 [129]. Wang et al. [130] reported that the methane reduction strategy through diet manipulation has its own benefit and drawbacks; therefore, further research is required (**Table 3**).

5. Tannin and saponin utilization to reduce methane emission

With the benefits of efficiency, plant extracts and their secondary metabolites have a high potential for ruminant methane mitigation. Incorporating saponin as ruminant feed can potentially suppress the production of methane, one of the biggest contributors to global warming [132]. Adding plant tannins to ruminant dietary can help mitigating methane emissions by reducing methanogenesis in rumen. This may be related to the antibacterial qualities of the tannins by decreasing fiber digestion and causing the ruminal microbial bacteria to not fully digest the feed [133]. Plant tannins, as feed supplements or as tanniferous forage diets, have shown a potential for reducing enteric CH_4 emissions by up to 20% [134, 135]. Many studies have assessed, both *in vitro* and *in vivo*, the connection between tannin-rich diets and ruminal CH_4 formation. The CH_4 reaction to tannin administration varies greatly based on the origin, variety, and molecular weight of the tannins, as well as the methanogenic ecosystem prevalent in the animal [84]. Tannins have anti-methanogenic ability, which has been demonstrated through *in vitro* evaluation. They can do this directly by suppressing methanogens or indirectly by affecting protozoa inside the rumen [136, 137]. Adding tannin to ruminant feed, either a tannin-containing diet or tannin extracts, can reduce enteric methane production [52]. A recent report using *in vivo* and *in vitro* methods assessed by Zhang et al. [118] showed that the addition of 30 and 60 g/kg of hydrolysable tannins (HT) to ruminant dietary was able to significantly reduce rumen CH_4 production by 37.6 and 36.4%, respectively. The effects of tannin addition in ruminant feed and its effect on methane mitigation are summarized in **Table 4**.

Tannin origin	Diets	Methane reduction effect	References
Rain tree pod meal (6 g/kg of total DMI)	Total mixed ration (concentrate + rice straw treated with urea) at 25 g/kg BW	10%	Anantasook et al. [138]
a. *Autocarpus integrifolia* leaf (186 g/kg DM) of CT b. *Ficus religiosa* leaf (13.5 g/kg DM) of HT c. *Jatropha curcus* (5.6 g/kg DM) of HT d. *Sesbania grandiflora* (13.1 g/kg DM) of HT	Elusine coracana straw and commercial concentrate mixture in 1:1 ratio	a. 4.73 (mL/total gas reduction) b. (mL/total gas reduction) c. (mL/total gas reduction) d. 2.02 (mL/total gas reduction)	Bhatta et al. [139]
Acacia (*Acacia molissima*) tannin extract	Forages (600–800 g/kg) and concentrates (200–400 g/kg)	a. Goat (13%) b. Sheep (23%) c. Buffalo (22%) d. Cattle (9%)	Bueno et al. [140]
a. *Acacia mearnsil* extract (82% CT) b. *Schinopsis balansae* extract (90.4% CT) c. *Castanea sativa* extract (5.7% CT and 75.5% HT) d. *Quercus aegilops* extract (8.0% CT and 71.2% HT)	Total mixed ration (forage / concentrate)	a. 12%, 21%, 32%, and 38% b. NE, 23%, 34%, and 40% c. 13%, 23%, 31%, and 40% d. 11%, 19%, 26%, and 36%	Hassanat and Benchaar [141]
Sainfoin (*Onobrychis viciifolia*) accessions: a. Rees "A" b. CPI63763 c. Cotswold Common d. CPI63767	50 mg lucerne (tannin free) / 30 ml of inoculum	a. 30% b. 45% c. 30% d. 48%	Hatew et al. [142]
a. Chestnut b. Sumac c. Mimosa d. Quebracho	380 mg (concentrate + hay) (30:70) / 30 mL of inoculum	a. 23% b. 30% c. 23% d. 27%	Jayanegara et al. [137]
a. *Trigonella foenumgraecum* leaf b. *Sesbania sesban* leaf	Hay: concentrate (50:50)	2. 20%	Jayanegara et al. [143]
a. Purified chestnut b. Sumac	Hay: concentrate (70:30)	a. 6.5% b. 7.2%	Jayanegara et al. [144]
a. Panicled-tick clover (PCT) b. *Sericea lespedeza* (SL)	Alfalfa: corn	a. 65% b. 24,4%	Naumann et al. [145]

Tannin origin	Diets	Methane reduction effect	References
Quebracho condensed tannin extract (75–77% QCT)	Corn: alfalfa	Ns, ns, ns	Pinski et al. [146]
Acacia cyanopylla (CT 63%)	Dates by-products and the vetch-oat	56.25% and 36.50%	Rira et al. [147]
Leucaena		a. 41.4 mL/g TDOM b. 47.4 (−14%) 1/kg DOM	Soltan et al. [148]
a. *Acacia saligna* leaves (6.3% CT) b. *Leucaena leucochepala leaves* (4.6% CT) c. *prosopis julifara* leaves (0.04% CT) d. *atriplex halimus leaves* (0.02% CT)	a. *Acasia saligna* b. *Leucaena leucochepala* c. *Prosopis julifora* d. *Atriplex halimus*	a. 38% b. 36% c. NE d. NE	Soltan et al. [149]
Leucaena leucochepala extract (100% CT) 10,15, 20, 25, and 30 mg	Guinea grass	−33%, −47%, −57%, −59%, and − 63%, respectively	Tan et al. [150]
Mangosteen peel powder		7%	Wanapat et al. [151]
a. Chestnut (*castaena sativa*) b. Valonea (*quercus valonea*)	Grass silage (100%)	a. 63% b. 34%	Wischer et al. [152]

DMI: dry matter intake, DM: dry matter, NA: not applicable, NE: no effect, ns: not significant, −: decrease compared to control, BW: body weight, CT: condensed tannin, HT: hydrolysable tannin, TDOM: truly degraded organic matter, DOM: degraded organic matter.

Table 4.
Effect of tannin addition on methane emission.

Adding saponin extract can also reduce methane emissions produced by ruminants, such as sheep and cattle (dairy and beef). It has been demonstrated that the extract from the leaf of *Sesbania sesban* or lucerne roots' saponins can significantly lower protozoa populations [67, 153, 154], which are crucial for the protein degradation of ruminal feed [155]. It is going to be difficult to determine the ideal doses of saponins to have a beneficial effect on rumen fermentation or ruminant production because saponins are typically supplied as extracts or as ground materials [156]. Very recent research conducted by Zhang et al. [118] reported that the addition of tea saponin extracts (5 g, 10 g, 20 g/kg DMI) was able to significantly reduce methane (CH_4) by 6.17 L, 7.86 L, and 10.53 L/kg DMI, respectively. Tannins and saponins extracts are recently available in the commercial market, and the newly developed hydrolysable tannins and tea saponin products are commercially available at very affordable prices (Biolink Biotechnology, Co, LTD, Beijing). The lowest market prices for these products are $11/kg (purity>81%) for hydrolysable tannins and $11/kg (purity>65%) for tea saponin products. When applying phytochemicals as feed additives, the amount and purity should be carefully monitored, as they may have anti-nutritional properties in larger quantities [131]. The effects of saponin addition in ruminant feed and its effect on methane mitigation are summarized in **Table 5**.

Saponin origin	Diets	Methane reduction effect	Reference
Purified saponin (1.55, 3.10, 4.65, and 6.20 mg/30 mL rumen inoculum)	Hybrid cumbu Napier grass	14.04, 21.90, 34.30, and 37.60%	Bharathidhasan et al. [157]
Papaya leaf (7.5, 12.5, and 25% of diet)	Concentrate (50%) + alfalfa (50%)	17, 34, and 37%	Jafari et al. [158, 159]
Papaya leaf methanol extract (PLE; 5, 10, and 15 mg of PLE/0.25 g DM)	Concentrate (50%) + alfalfa (50%)	Ns, ns, and 34%	Jafari et al. [158, 159]
Papaya leaf solvent fractions (PLF; 15 mg of PLF/0.25 g DM)	Concentrate (50%) + alfalfa (50%)	25%, 29%, ns, 25% and ns	Jafari et al. [160]
Yucca saponin (8.5% saponin)	Total mixed ration (forage/concentrate)	NA	Li and Powers [161]
Yucca schidigera	Forage and concentrate (65:35)	15%	Narvaez et al. [162]
a. Quillaja saponin (0.6 g/L) b. Quillaja saponin (1.2 g/L) c. Quillaja saponin (1.2 g/L) + propionic acid (8 mM) + nitrate (10 mM)	Corn silage (45%) + alfalfa hay (10%) + dairy protein product (20%) + concentrate mixture (25%)	a. 11% b. 24% c. 85%	Patra and Yu [163]
a. Quillaja saponin (0.6 g/L) b. Quillaja saponin (0.6 g/L) + nitrate (5 mM) and sulfate (5 mM)	Corn silage (45%) + alfalfa hay (10%) + dairy protein product (20%) + concentrate mixture (25%)	a. 8% b. 47%	Patra and Yu [164]
a. Quillaja saponin b. Saponin + garlic c. Saponin + nitrate d. Saponin + garlic + nitrate	Concentrate and alfalfa (70:30)	a. 36% b. 45% c. 55% d. 70%	Patra and Yu [165]
Yucca schidigera (4.4% saponin)	Dates by-product + the vetch + oat	60%	Rira et al. [147]
Mangosteen peel powder (10.9% saponin)	Concentrate + rice straw	7%	Wanapat et al. [151]

DM: dry matter, NA: not applicable, ns: not significant.

Table 5.
Effect of saponin addition on methane emission.

6. Summary, conclusion, and future study

Based on the scientific findings presented in this chapter, the following most important conclusions can be drawn:

1. Canadian Prairie region has an abundant amount of unused products with affordable prices that can possibly be used as a source of ruminant feed that offers high energy and starch contents, but it is not recommended to be used directly without processing.

2. It is founded that pellet processing of blended feed substances can balance amino acid delivery, enhance and optimize nutritional supply, and alter rumen fermentation behavior in ruminants and can also reinforce technical advantages including enhanced stability, simpler handling, and storage.

3. Greenhouse gas (GHG) emission has become a joint challenge in the last few decades and numerous attempts have been made to reduce CH_4 production in ruminants with different approaches. However, those approaches still have some drawbacks (e.g., costly, resistant, residue, toxic) which detain its effectivity and application.

4. Tannin and saponin are two phytochemicals derived from plant materials with approved methane reduction agents. However, there is no literature on the effects of unused prairie products combined with those newly developed hydrolysable tannins (or saponins) at different levels. Therefore, a further investigation is necessary to study the effect of pellet processing of this combination on (1) bioactive compound (CT) levels, (2) amino acid profile, (3) physiochemical and nutrient profiles, (4) nutrient fermentation on GHG emission, utilization, and availability in rumen and intestine in ruminants, (5) protein and energy metabolic characteristics and truly absorbed nutrient supply in ruminant system, (6) changes on molecular structure in relation to nutrient utilization availability, and (7) animal metabolic characteristics and production performance.

Acknowledgements

The Ministry of Agriculture Strategic Research Chair (PY) Research Programs have been financially supported by various grants from the Natural Sciences and Engineering Research Council of Canada (NSERC-Individual Discovery Grant and NSERC-CRD Grant), Saskatchewan Agriculture Strategic Feed Research Program Fund, Agricultural Development Fund (ADF), SaskMilk, SaskCanola, AlbertaMilk, Saskatchewan Forage Network (SNK), Western Grain Research Foundation (WGRF), SaskPulse Growers, Prairie Oat Growers Association (POGA), etc. The JH research programs are supported by Inner Mongolia Autonomous Region Science and Technology Plan Project, National Foreign Experts Introduction Project, Technical Innovation System of Sheep Industry in Inner Mongolia, and Ordos Major Project in Inner Mongolia. The Feed Molecular Structure Research Programs have been supported by Canadian Light Sources (CLS, University of Saskatchewan, Canada), National Synchrotron Light Sources- Brookhaven National Lab (NSLS-BNL, New York, U.S. Department of Energy, USA), Advanced Light Source-Berkeley Lawrence National Lab (ALS -BLNL, Berkeley, U.S. Department of Energy, USA).

Author details

Taufiq Hidayat[1], Maria Eugenia Rodriguez Espinosa[1], Xiaogang Yan[1,2], Katerina Theodoridou[1,3], Samadi[1,4], Quanhui Peng[1,5], Bin Feng[5], Weixian Zhang[6], Jiangfeng He[7] and Peiqiang Yu[1*]

1 Department of Animal and Poultry Science, College of Agriculture and Bioresources, University of Saskatchewan, Canada

2 Institute of Animal Nutrition and Feed Science, Jilin Academy of Agricultural Sciences, Gongzhuling City, China

3 Institute for Global Food Security, Queen's University Belfast, United Kingdom

4 Agricultural Faculty, Animal Husbandry Department, Universitas Syiah Kuala, Darussalam - Banda Aceh, Indonesia

5 Animal Nutrition Institute, Sichuan Agricultural University, Chengdu, Sichuan, China

6 Henan University of Animal Husbandry and Economy, Zhengzhou, China

7 Institute of Biotechnology, Inner Mongolia Academy of Agriculture and Husbandry Science, Hohhot, China

*Address all correspondence to: peiqiang.yu@usask.ca

References

[1] Environment Canada. The State of Canada's Environment, Government of Canada. Ottawa; 1996. Available from: https://publications.gc.ca/collections/collection_2019/eccc/En21-54-1996-eng.pdf

[2] National Wetlands Working Group. Wetlands of Canada (ecological land classification series, No. 24). Journal of Environmental Quality. 1998;**19**:350. Available from: https://publications.gc.ca/site/eng/9.867553/publication.html

[3] Statistics Canada. Seeding decisions harvest opportunities for Canadian farmers. 2017. Available from: https://www150.statcan.gc.ca/n1/pub/95-640-x/2016001/article/14813-eng.htm

[4] The Canada Guide. The Prairies. Canada. Available from: https://thecanadaguide.com/places/the-prairies/; n.d. [Accessed: May 28, 2023]

[5] Canada Agriculture Census. Canada's 2021 Census of Agriculture: A Closer Look at Farming across the Regions. Statistics Canada: Canada Agriculture Census; 2021. Available from https://www150.statcan.gc.ca/n1/en/daily-quotidien/220615/dq220615a-eng.pdf?st=1L-8mSLU

[6] Mupondwa E, Li X, Tabil L, Sokhansanj S, Adapa P. Status of Canada's lignocellulosic ethanol: Part I: Pretreatment technologies. Renewable and Sustainable Energy Reviews. 2017;**72**:178-190. DOI: 10.1016/j.rser.2017.01.039

[7] Patel M, Oyedun AO, Kumar A, Gupta R. What is the production cost of renewable diesel from woody biomass and agricultural residue based on experimentation? A comparative assessment. Fuel Processing Technology. 2019;**191**:79-92. DOI: 10.1016/j.fuproc.2019.03.026

[8] Shabani N, Sowlati T. A mixed integer non-linear programming model for tactical value chain optimization of a wood biomass power plant. Applied Energy. 2013;**104**:353-361. DOI: 10.1016/j.apenergy.2012.11.013

[9] NDSU. Feeding Field Peas to Livestock. 2002. Available from: https://www.ag.ndsu.edu/pubs/ansci/livestoc/as.1224.Pdf.

[10] Nasir G, Zaidi S, Tabassum N, Asfaq. A review on nutritional composition, health benefits and potential applications of by-products from pea processing. Biomass Conversion and Biorefinery. 2022;**6**:1-14. DOI: 10.1007/s13399-022-03324-0

[11] Perz K, Kaczmarek SA, Nowaczewski S, Cowieson AJ, Jarosz Ł, Ciszewski A, et al. The effect of reduction of resistant starch content of faba bean and pea by amylase supplementation on performance, nutrient digestibility, and sialic acid excretion of broiler chickens. Animal Feed Science and Technology. 2023;**298**:115621. DOI: 10.1016/j.anifeedsci.2023.115621

[12] Heendeniya RG, Christensen DA, Maenz DD, McKinnon JJ, Yu P. Protein fractionation by-product from canola meal for dairy cattle. Journal of Dairy Science. 2012;**95**:4488-4500

[13] Hiclking D. Maximized utilization of canola co-products in the livestock industry. In: 29th Western Nutrition Conference. Edmonton, Canada: University of Alberta; 2008. pp. 23-24

[14] Sekali M, Mlambo V, Marume U, Mathuthu M. *In vitro* ruminal

fermentation parameters of canola meal protein in response to incremental doses of gamma irradiation. South African Journal of Animal Science. 2023;**53**(1):109-116. DOI: 10.4314/sajas.v53i1.12

[15] Newkirk R. Canola Meal: Feed Industry Guide. 4th ed. Winnipeg, Manitoba: Canadian International Grains Institute; 2009. pp. 3-14. Available from: https://cigi.ca/wp-content/uploads/2011/12/2009-Canola_Guide.pdf

[16] Wright CF, von Keyserlingk MAG, Swift ML, Fisher LJ, Shelford JA, Dinn NE. Heat- and Lignosulfonate-treated canola meal as a source of ruminal undegradable protein for lactating dairy cows. Journal of Dairy Science. 2005;**88**(1):238-243. DOI: 10.3168/jds.S0022-0302(05)72681-3

[17] Babu S, Singh Rathore S, Singh R, Kumar S, Singh VK, Yadav SK, et al. Exploring agricultural waste biomass for energy, food and feed production and pollution mitigation: A review. Bioresource Technology. 2022;**360**:127566. DOI: 10.1016/j.biortech.2022.127566

[18] Poore J, Nemecek T. Reducing food's environmental impacts through producers and consumers. Science. 2018;**360**(6392):987-992. DOI: 10.1126/science.aaq0216

[19] Halmemies-Beauchet-Filleau A, Rinne M, Lamminen M, Mapato C, Ampapon T, Wanapat M, et al. Review: Alternative and novel feeds for ruminants: Nutritive value, product quality and environmental aspects. Animal. 2018;**12**:s295-s309. DOI: 10.1017/S1751731118002252

[20] Chen X, Zhao X, Ge J, Zhao Y, Wei Z, Yao C, et al. Recognition of the neutral sugars conversion induced by bacterial community during lignocellulose wastes composting. Bioresource Technology. 2019;**294**:122153. DOI: 10.1016/j.biortech.2019.122153

[21] Wang K, He C, You S, Liu W, Wang W, Zhang R, et al. Transformation of organic matters in animal wastes during composting. Journal of Hazardous Materials. 2015;**300**:745-753. DOI: 10.1016/j.jhazmat.2015.08.016

[22] Wang R, Zhao Y, Xie X, Mohamed TA, Zhu L, Tang Y, et al. Role of NH_3 recycling on nitrogen fractions during sludge composting. Bioresource Technology. 2020;**295**:122175. DOI: 10.1016/j.biortech.2019.122175

[23] Regulation EC No. 1831/2003. Regulation (EC) No. 1831/2003 of the European Parliament and of the council on additives for use in animal nutrition. Official Journal L 268. 2003a, 2003;**268**:29-43. European Union. Available from: https://faolex.fao.org/docs/pdf/eur40306original.pdf

[24] Cieslak A, Szumacher-Strabel M, Stochmal A, Oleszek W. Plant components with specific activities against rumen methanogens. Animal. 2013;7:253-265. DOI: 10.1017/S1751731113000852

[25] Makkar HPS, Francis G, Becker K. Bioactivity of phytochemicals in some lesser-known plants and their effects and potential applications in livestock and aquaculture production systems. Animal. 2007;**1**(9):1371-1391. DOI: 10.1017/S1751731107000298

[26] Petersson-Wolfe CS, Leslie KE, Osborne T, McBride BW, Bagg R, Vessie G, et al. Effect of monensin delivery method on dry matter intake, body condition score, and metabolic parameters in transition dairy cows. Journal of Dairy Science. 2007;**90**(4):1870-1879. DOI: 10.3168/jds.2006-402

[27] Ahmed H, Jahan S, Khan A, Khan L, Khan BT, Ullah H, et al. Supplementation of green tea extract (GTE) in extender improves structural and functional characteristics, total antioxidant capacity and in vivo fertility of buffalo (*Bubalus bubalis*) bull spermatozoa. Theriogenology. 2020;**145**:190-197. DOI: 10.1016/j.theriogenology.2019.10.024

[28] Miller-Webster T, Hoover WH, Holt M, Nocek JE. Influence of yeast culture on ruminal microbial metabolism in continuous culture. Journal of Dairy Science. 2002;**85**(8):2009-2014. DOI: 10.3168/jds.S0022-0302(02)74277-X

[29] Alemu A, Romero-Pérez A, Araujo R, Beauchemin K. Effect of encapsulated nitrate and microencapsulated blend of essential oils on growth performance and methane emissions from beef steers fed backgrounding diets. Animals. 2019;**9**(1):21. DOI: 10.3390/ani9010021

[30] Castillo-Castillo Y, Ruiz-Barrera O, Burrola-Barraza ME, Marrero-Rodriguez Y, Salinas-Chavira J, Angulo-Montoya C, et al. Isolation and characterization of yeasts from fermented apple bagasse as additives for ruminant feeding. Brazilian Journal of Microbiology. 2016;**47**(4):889-895. DOI: 10.1016/j.bjm.2016.07.020

[31] Dann HM, Drackley JK, McCoy GC, Hutjens MF, Garrett JE. Effects of yeast culture (*Saccharomyces cerevisiae*) on prepartum intake and postpartum intake and milk production of Jersey cows. Journal of Dairy Science. 2000;**83**(1):123-127. DOI: 10.3168/jds.S0022-0302(00)74863-6

[32] Lila ZA, Mohammed N, Yasui T, Kurokawa Y, Kanda S, Itabashi H. Effects of a twin strain of *Saccharomyces cerevisiae* live cells on mixed ruminal microorganism fermentation in vitro. Journal of Animal Science. 2004;**82**(6):1847-1854. DOI: 10.2527/2004.8261847x

[33] Beauchemin KA, McGinn SM. Methane emissions from beef cattle: Effects of fumaric acid, essential oil, and canola oil 1. Journal of Animal Science. 2006;**84**(6):1489-1496. DOI: 10.2527/2006.8461489x

[34] Beauchemin KA, McGinn SM, Petit HV. Methane abatement strategies for cattle: Lipid supplementation of diets. Canadian Journal of Animal Science. 2007;**87**(3):431-440. DOI: 10.4141/CJAS07011

[35] Carulla JE, Kreuzer M, Machmüller A, Hess HD. Supplementation of Acacia mearnsii tannins decreases methanogenesis and urinary nitrogen in forage-fed sheep. Australian Journal of Agricultural Research. 2005;**56**(9):961. DOI: 10.1071/AR05022

[36] Beeman K. The effect of lactobacillus spp. on convalescing calves. In: Agri Practice. Vol. 6. CABI Record Number: 19862276303. CABI Digital Library; 1985. ISSN (Print): 0745-452X

[37] Gilliland SE, Bruce BB, Bush LJ, Staley TE. Comparisons of two strains of *Lactobacillus acidophilus* as dietary adjuncts for young calves. Journal of Dairy Science. 1980;**63**(6):964-972. DOI: 10.3168/jds.S0022-0302(80)83033-5

[38] Lee RW, Botts RL. Evaluation of single oral dosing and continuous feeding of *Streptococcus faecium* M74 (Syntabac) on performance of incoming feedlot cattle. Journal of Animal Science. 1988;**66**(suppl. 1):460

[39] Pond KR, Goode L. The evaluation of Probios® for wean-stressed lambs. Pond, K. Microbial Genetics Division.

Des Moines, Iowa: Pioneer Hi-Bred International; 1985

[40] Umberger SH, Notter DR, Webb KE, McClure WH. Evaluation of a lactobacillus inoculant on feedlot lamb performance. Journal of Animal Science. 1989;**8**:40-45

[41] Ahmed E, Batbekh B, Fukuma N, Kand D, Hanada M, Nishida T. A garlic and citrus extract: Impacts on behavior, feed intake, rumen fermentation, and digestibility in sheep. Animal Feed Science and Technology. 2021;**278**:115007. DOI: 10.1016/j.anifeedsci.2021.115007

[42] Herrick KJ, Hippen AR, Kalscheur KF, Anderson JL, Ranathunga SD, Patton RS, et al. Lactation performance and digestibility of forages and diets in dairy cows fed a hemicellulose extract. Journal of Dairy Science. 2012;**95**(6):3342-3353. DOI: 10.3168/jds.2011-5168

[43] Lamminen M, Halmemies-Beauchet-Filleau A, Kokkonen T, Simpura I, Jaakkola S, Vanhatalo A. Comparison of microalgae and rapeseed meal as supplementary protein in the grass silage based nutrition of dairy cows. Animal Feed Science and Technology. 2017;**234**:295-311. DOI: 10.1016/j.anifeedsci.2017.10.002

[44] Orzuna-Orzuna JF, Dorantes-Iturbide G, Lara-Bueno A, Mendoza-Martínez GD, Miranda-Romero LA, Hernández-García PA. Effects of dietary tannins' supplementation on growth performance, rumen fermentation, and enteric methane emissions in beef cattle: A meta-analysis. Sustainability. 2021;**13**(13):7410

[45] Dai X, Faciola AP. Evaluating strategies to reduce ruminal protozoa and their impacts on nutrient utilization and animal performance in ruminants – A meta-analysis. Frontiers in Microbiology. 2019;**10**:2648. DOI: 10.3389/fmicb.2019.02648

[46] Choi YY, Lee SJ, Kim HS, Eom JS, Kim DH, Lee SS. The potential nutritive value of *Sargassum fulvellum* as a feed ingredient for ruminants. Algal Research. 2020;**45**:101761. DOI: 10.1016/j.algal.2019.101761

[47] Mueller-Harvey I. Unravelling the conundrum of tannins in animal nutrition and health. Journal of the Science of Food and Agriculture. 2006;**86**(13):2010-2037. DOI: 10.1002/jsfa.2577

[48] Goel G, Puniya AK, Aguilar CN, Singh K. Interaction of gut microflora with tannins in feeds. Naturwissenschaften. 2005;**92**(11):497-503. DOI: 10.1007/s00114-005-0040-7

[49] Reed JD. Nutritional toxicology of tannins and related polyphenols in forage legumes. Journal of Animal Science. 1995;**73**(5):1516-1528. DOI: 10.2527/1995.7351516x

[50] Makkar HPS. Effects and fate of tannins in ruminant animals, adaptation to tannins, and strategies to overcome detrimental effects of feeding tannin-rich feeds. Small Ruminant Research. 2003;**49**(3):241-256. DOI: 10.1016/S0921-4488(03)00142-1

[51] Tiemann TT, Lascano CE, Wettstein H-R, Mayer AC, Kreuzer M, Hess HD. Effect of the tropical tannin-rich shrub legumes *Calliandra calothyrsus* and *Flemingia macrophylla* on methane emission and nitrogen and energy balance in growing lambs. Animal. 2008;**2**(5):790-799. DOI: 10.1017/S1751731108001791

[52] Jayanegara A, Leiber F, Kreuzer M. Meta-analysis of the

relationship between dietary tannin level and methane formation in ruminants from *in vivo* and *in vitro* experiments. Journal of Animal Physiology and Animal Nutrition. 2012;**96**(3):365-375. DOI: 10.1111/j.1439-0396.2011.01172.x

[53] McSweeney CS, Palmer B, McNeill DM, Krause DO. Microbial interactions with tannins: Nutritional consequences for ruminants. Animal Feed Science and Technology. 2001;**91**(1-2):83-93. DOI: 10.1016/S0377-8401(01)00232-2

[54] Stewart AJ, Stewart RF. Phenols. In: Encyclopedia of Ecology. Amsterdam, Netherlands: Elsevier; 2008. pp. 2682-2689. DOI: 10.1016/B978-008045405-4.00417-1

[55] Hartzfeld PW, Forkner R, Hunter MD, Hagerman AE. Determination of hydrolyzable tannins (Gallotannins and Ellagitannins) after reaction with potassium iodate. Journal of Agricultural and Food Chemistry. 2002;**50**(7):1785-1790. DOI: 10.1021/jf0111155

[56] Zhong RZ, Sun HX, Liu HW, Zhou DW. Effects of tannic acid on *Haemonchus contortus* larvae viability and immune responses of sheep white blood cells *in vitro*. Parasite Immunology. 2014;**36**(2):100-106. DOI: 10.1111/pim.12092

[57] Sparg SG, Light ME, van Staden J. Biological activities and distribution of plant saponins. Journal of Ethnopharmacology. 2004;**94**(2-3, 243):219. DOI: 10.1016/j.jep.2004.05.016

[58] Osbourn A. Saponins and plant defence — A soap story. Trends in Plant Science. 1996;**1**(1):4-9. DOI: 10.1016/S1360-1385(96)80016-1

[59] Osbourn AE. Saponins in cereals. Phytochemistry. 2003;**62**(1):1-4. DOI: 10.1016/S0031-9422(02)00393-X

[60] Rupasinghe HPV, Jackson C-JC, Poysa V, Di Berardo C, Bewley JD, Jenkinson J. Soyasapogenol A and B distribution in soybean (Glycine max L. Merr.) in relation to seed physiology, genetic variability, and growing location. Journal of Agricultural and Food Chemistry. 2003;**51**(20):5888-5894. DOI: 10.1021/jf0343736

[61] El Aziz MMA, Ashour AS, Melad AG. A review on saponins from medicinal plants: Chemistry, isolation, and determination. International Journal of Nanomedicine. 2019;**8**(1):282-288

[62] Güçlü-Üstündağ Ö, Mazza G. Saponins: Properties, applications and processing. Critical Reviews in Food Science and Nutrition. 2007;**47**(3):231-258. DOI: 10.1080/10408390600698197

[63] Madland E. Extraction, Isolation and Structure Elucidation of Saponins from Herniaria incana [Thesis]. Norway: Norwegian University of Science and Technology; 2013

[64] Reddy NV, Ghosh SB, Bobbu P, Anitha D, Tartte V. Triterpenoid saponins: A review on biosynthesis, applications and mechanism of their action. International Journal of Pharmacy and Pharmaceutical Sciences. 2015;**7**(1):24-28

[65] Francis G, Kerem Z, Makkar HPS, Becker K. The biological action of saponins in animal systems: A review. British Journal of Nutrition. 2002;**88**(6):587-605. DOI: 10.1079/BJN2002725

[66] Das T, Banerjee D, Chakraborty D, Pakhira M, Shrivastava B, Kuhad R. Saponin: Role in animal system. Veterinary World. 2012;**5**(4):248. DOI: 10.5455/vetworld.2012.248-254

[67] Newbold CJ, El Hassan SM, Wang J, Ortega ME, Wallace RJ. Influence of foliage

from African multipurpose trees on activity of rumen protozoa and bacteria. British Journal of Nutrition. 1997a;**78**(2):237-249. DOI: 10.1079/BJN19970143

[68] Wallace RJ, Arthaud L, Newbold CJ. Influence of *Yucca shidigera* extract on ruminal ammonia concentrations and ruminal microorganisms. Applied and Environmental Microbiology. 1994;**60**(6):1762-1767. DOI: 10.1128/aem.60.6.1762-1767.1994

[69] Rakic L. Feed Structure: Effects on Physical Quality of Feed, Chemical Status of the Feed and Nutritional Consequence. Ås, Norway: Norwegian University of Life Science; 2012

[70] Gustafson ML. The durability test – a key to handling wafers and pellets. ASAE Paper No. 59621. St. Joseph, Mich.: ASAE; 1959

[71] Bonfante E, Palmonari A, Mammi L, Canestrari G, Fustini M, Formigoni A. Effects of a completely pelleted diet on growth performance in Holstein heifers. Journal of Dairy Science. 2016;**99**(12):9724-9731. DOI: 10.3168/jds.2016-11033

[72] Khafipour E, Krause DO, Plaizier JC. Alfalfa pellet-induced subacute ruminal acidosis in dairy cows increases bacterial endotoxin in the rumen without causing inflammation. Journal of Dairy Science. 2009;**92**(4):1712-1724. DOI: 10.3168/jds.2008-1656

[73] Miron J, Yosef E, Ben-Ghedalia D. Composition and *in vitro* digestibility of monosaccharide constituents of selected byproduct feeds. Journal of Agricultural and Food Chemistry. 2001;**49**(5):2322-2326. DOI: 10.1021/jf0008700

[74] Thomas M, van der Poel AFB. Physical quality of pelleted animal feed 1. Criteria for pellet quality. Animal Feed Science and Technology. 1996;**61**(1-4):89-112. DOI: 10.1016/0377-8401(96)00949-2

[75] Johnson KA, Johnson DE. Methane emissions from cattle. Journal of Animal Science. 1995;**73**(8):2483-2492. DOI: 10.2527/1995.7382483x

[76] McGeough EJ, Passetti LCG, Chung YH, Beauchemin KA, McGinn SM, Harstad OM, et al. Methane emissions, feed intake, and total tract digestibility in lambs fed diets differing in fat content and fibre digestibility. Canadian Journal of Animal Science. 2019;**99**(4):858-866. DOI: 10.1139/cjas-2018-0185

[77] Britannica. Methane. Encyclopedia Britannica. n.d. Available from: https://www.britannica.com/science/methane [Accessed: June 3, 2023]

[78] Martin C, Doreau M, Morgavi DP. Methane mitigation in ruminants: From rumen microbes to the animal. In: Rowlinson P, Steele M, Nefzaoui A, editors. Livestock an Global Climate Change. Cambridge, UK: Cambridge University Press; 2008. pp. 130-133

[79] Janssen PH, Kirs M. Structure of the archaeal community of the Rumen. Applied and Environmental Microbiology. 2008;**74**(12):3619-3625. DOI: 10.1128/AEM.02812-07

[80] Ferry JG, Kasted KA. Methanogenesis. In: Cavicchioli R, editor. Archaea: Molecular Cell Biology. Washington, D.C: ASM Press; 2007. pp. 288-314

[81] Guzman CE, Bereza-Malcolm LT, De Groef B, Franks AE. Presence of selected methanogens, fibrolytic bacteria, and proteobacteria in the gastrointestinal tract of neonatal dairy calves from birth to 72 hours. PLoS One. 2015;**10**(7):e0133048. DOI: 10.1371/journal.pone.0133048

[82] Skillman LC, Evans PN, Naylor GE, Morvan B, Jarvis GN, Joblin KN. 16S ribosomal DNA-directed PCR primers for ruminal methanogens and identification of methanogens colonising young lambs. Anaerobe. 2004;**10**(5):277-285. DOI: 10.1016/j.anaerobe.2004.05.003

[83] Liu Y, Whitman WB. Metabolic, phylogenetic, and ecological diversity of the methanogenic archaea. Annals of the New York Academy of Sciences. 2008;**1125**(1):171-189. DOI: 10.1196/annals.1419.019

[84] Beauchemin KA, Ungerfeld EM, Eckard RJ, Wang M. Review: Fifty years of research on rumen methanogenesis: Lessons learned and future challenges for mitigation. Animal. 2020a;**14**:s2-s16. DOI: 10.1017/S1751731119003100

[85] Morgavi DP, Cantalapiedra-Hijar G, Eugène M, Martin C, Noziere P, Popova M, et al. Review: Reducing enteric methane emissions improves energy metabolism in livestock: Is the tenet right? Animal. 2023;**17 Suppl 3**:100830. DOI: 10.1016/j.animal.2023.100830

[86] Martin C, Morgavi DP, Doreau M. Methane mitigation in ruminants: From microbe to the farm scale. Animal. 2010;**4**(3):351-365. DOI: 10.1017/S1751731109990620

[87] McAllister TA, Meale SJ, Valle E, Guan LL, Zhou M, Kelly WJ, et al. Ruminant nutrition symposium: Use of genomics and transcriptomics to identify strategies to lower ruminal methanogenesis 1, 2, 3. Journal of Animal Science. 2015;**93**(4):1431-1449. DOI: 10.2527/jas.2014-8329

[88] Kittelmann S, Seedorf H, Walters WA, Clemente JC, Knight R, Gordon JI, et al. Simultaneous amplicon sequencing to explore co-occurrence patterns of bacterial, archaeal and eukaryotic microorganisms in rumen microbial communities. PLoS One. 2013;**8**(2):e47879. DOI: 10.1371/journal.pone.0047879

[89] Neill AR, Grime DW, Dawson RMC. Conversion of choline methyl groups through trimethylamine into methane in the rumen. Biochemical Journal. 1978;**170**(3):529-535. DOI: 10.1042/bj1700529

[90] Poulsen M, Schwab C, Borg Jensen B, Engberg RM, Spang A, Canibe N, et al. Methylotrophic methanogenic thermoplasmata implicated in reduced methane emissions from bovine rumen. Nature Communications. 2013;**4**(1):1428. DOI: 10.1038/ncomms2432

[91] Morgavi DP, Forano E, Martin C, Newbold CJ. Microbial ecosystem and methanogenesis in ruminants. Animal. 2010;**4**(7):1024-1036. DOI: 10.1017/S1751731110000546

[92] Nur K, Atabany A, Muladno, Jayanegara A. Methan production of dairy cow ruminants with different feed and effect on the production and quality of milk. Jurnal Ilmu Produksi Dan Teknologi Hasil Peternakan. 2015;**3**(2):65-71

[93] Broucek J. Production of methane emissions from ruminant husbandry: A review. Journal of Environmental Protection. 2014;**05**(15):1482-1493. DOI: 10.4236/jep.2014.515141

[94] Shibata M, Terada F. Factors affecting methane production and mitigation in ruminants. Animal Science Journal. 2010;**81**(1):2-10. DOI: 10.1111/j.1740-0929.2009.00687.x

[95] Sauvant D, Giger-Reverdin S. In: Ortigues-Marty I, Miraux N,

Brand-Williams W, editors. Energy and Protein Metabolism and Nutrition. Vol. 124. Wageningen, Netherlands: Wageningen Academic Publishers; 2007. DOI: 10.3920/978-90-8686-613-7

[96] Olijhoek DW, Hellwing ALF, Noel SJ, Lund P, Larsen M, Weisbjerg MR, et al. Feeding up to 91% concentrate to Holstein and Jersey dairy cows: Effects on enteric methane emission, rumen fermentation and bacterial community, digestibility, production, and feeding behavior. Journal of Dairy Science. 2022;**105**(12):9523-9541. DOI: 10.3168/jds.2021-21676

[97] Plaizier JC, Krause DO, Gozho GN, McBride BW. Subacute ruminal acidosis in dairy cows: The physiological causes, incidence and consequences. The Veterinary Journal. 2008;**176**(1):21-31. DOI: 10.1016/j.tvjl.2007.12.016

[98] Charmley E, Williams SRO, Moate PJ, Hegarty RS, Herd RM, Oddy VH, et al. A universal equation to predict methane production of forage-fed cattle in Australia. Animal Production Science. 2016;**56**(3):169. DOI: 10.1071/AN15365

[99] Sun X, Cheng L, Jonker A, Munidasa S, Pacheco D. A review: Plant carbohydrate types—the potential impact on ruminant methane emissions. Frontiers in Veterinary Science. 2022;**9**:880115. pp 1-15. DOI: 10.3389/fvets.2022.880115

[100] Duthie C-A, Haskell M, Hyslop JJ, Waterhouse A, Wallace RJ, Roehe R, et al. The impact of divergent breed types and diets on methane emissions, rumen characteristics and performance of finishing beef cattle. Animal. 2017;**11**(10):1762-1771. DOI: 10.1017/S1751731117000301

[101] Tapio I, Snelling TJ, Strozzi F, Wallace RJ. The ruminal microbiome associated with methane emissions from ruminant livestock. Journal of Animal Science and Biotechnology. 2017;**8**(1):7. DOI: 10.1186/s40104-017-0141-0

[102] Danielsson R, Dicksved J, Sun L, Gonda H, Müller B, Schnürer A, et al. Methane production in dairy cows correlates with rumen methanogenic and bacterial community structure. Frontiers in Microbiology. 2017;**8**:226. DOI: 10.3389/fmicb.2017.00226

[103] Danielsson R, Schnürer A, Arthurson V, Bertilsson J. Methanogenic population and CH_4 production in Swedish dairy cows fed different levels of forage. Applied and Environmental Microbiology. 2012;**78**(17):6172-6179. DOI: 10.1128/AEM.00675-12

[104] Kittelmann S, Pinares-Patiño CS, Seedorf H, Kirk MR, Ganesh S, McEwan JC, et al. Two different bacterial community types are linked with the low-methane emission trait in sheep. PLoS One. 2014;**9**(7):e103171. DOI: 10.1371/journal.pone.0103171

[105] Morgavi DP, Martin C, Jouany J-P, Ranilla MJ. Rumen protozoa and methanogenesis: Not a simple cause–effect relationship. British Journal of Nutrition. 2012;**107**(3):388-397. DOI: 10.1017/S0007114511002935

[106] Shi W, Moon CD, Leahy SC, Kang D, Froula J, Kittelmann S, et al. Methane yield phenotypes linked to differential gene expression in the sheep rumen microbiome. Genome Research. 2014;**24**(9):1517-1525. DOI: 10.1101/gr.168245.113

[107] Wallace RJ, Rooke JA, Duthie C-A, Hyslop JJ, Ross DW, McKain N, et al. Archaeal abundance in post-mortem ruminal digesta may help predict methane emissions from beef cattle.

Scientific Reports. 2014;**4**(1):5892. DOI: 10.1038/srep05892

[108] Zhou M, Chung Y-H, Beauchemin KA, Holtshausen L, Oba M, McAllister TA, et al. Relationship between rumen methanogens and methane production in dairy cows fed diets supplemented with a feed enzyme additive. Journal of Applied Microbiology. 2011;**111**(5):1148-1158. DOI: 10.1111/j.1365-2672.2011.05126.x

[109] Kurihara M, Kume S, Aii T, Takahashi S, Shibata M, Nishida T. Feeding method for dairy cattle to cope with global warming – technical assessment based on energy metabolism. The Bulletin of the Kyushu National Agricultural Experiment Station. 1995;**29**:21-107

[110] Lee MA, Davis AP, Chagunda MGG, Manning P. Forage quality declines with rising temperatures, with implications for livestock production and methane emissions. Biogeosciences. 2017;**14**(6):1403-1417. DOI: 10.5194/bg-14-1403-2017

[111] Yadav B, Singh G, Wankar A, Dutta N, Chaturvedi VB, Verma MR. Effect of simulated heat stress on digestibility, methane emission and metabolic adaptability in crossbred cattle. Asian-Australasian Journal of Animal Sciences. 2016;**29**(11):1585-1592. DOI: 10.5713/ajas.15.0693

[112] McManus C, Paludo GR, Louvandini H, Gugel R, Sasaki LCB, Paiva SR. Heat tolerance in Brazilian sheep: Physiological and blood parameters. Tropical Animal Health and Production. 2009;**41**(1):95-101. DOI: 10.1007/s11250-008-9162-1

[113] Mitchell JFB. The "greenhouse" effect and climate change. Reviews of Geophysics. 1989;**27**(1):115. DOI: 10.1029/RG027i001p00115

[114] Moss AR, Jouany J-P, Newbold J. Methane production by ruminants: Its contribution to global warming. Annales de Zootechnie. 2000;**49**(3):231-253. DOI: 10.1051/animres:2000119

[115] Change C. IPCC fourth assessment report. The Physical Science Basis. 2007;**2**:580-595

[116] Ehhalt D, Prather M, Dentener F, Derwent R, Dlugokencky EJ, Holland EI et al. Atmospheric Chemistry and Greenhouse Gases. Houghton JT et al. Cambridge, United Kingdom: Cambridge University Pres; 2001. Available from: https://www. osti.gov/biblio/901482

[117] EPA. Overview of Greenhouse Gases. United States Environmental Protection Agency, Washington DC, USA: Climate Change Division, Office of Atmospheric Programs (MC-6207A). 2023. Available from: https://www.epa.gov/ghgemissions/overview-greenhouse-gases

[118] Zhang F, Li B, Ban Z, Liang H, Li L, Zhao W, et al. Evaluation of origanum oil, hydrolysable tannins and tea saponin in mitigating ruminant methane: *In vitro* and *in vivo* methods. Journal of Animal Physiology and Animal Nutrition. 2021;**105**(4):630-638. DOI: 10.1111/jpn.13501

[119] Tajima K, Aminov RI, Nagamine T, Matsui H, Nakamura M, Benno Y. Diet-dependent shifts in the bacterial population of the rumen revealed with real-time PCR. Applied and Environmental Microbiology. 2001;**67**(6):2766-2774. DOI: 10.1128/AEM.67.6.2766-2774.2001

[120] Wadhwa M, Bakshi MPS, Makkar HPS. Modifying gut microbiomes in large ruminants: Opportunities in non-intensive husbandry systems. Animal Frontiers. 2016;**6**(2):27. DOI: 10.2527/af.2016-0020

[121] Wright A-DG, Toovey AF, Pimm CL. Molecular identification of methanogenic archaea from sheep in Queensland, Australia reveal more uncultured novel archaea. Anaerobe. 2006;**12**(3):134-139. DOI: 10.1016/j.anaerobe.2006.02.002

[122] Palangi V, Taghizadeh A, Abachi S, Lackner M. Strategies to mitigate enteric methane emissions in ruminants: A review. Sustainability. 2022;**14**(20):13229. DOI: 10.3390/su142013229

[123] Beauchemin KA, Ungerfeld EM, Abdalla AL, Alvarez C, Arndt C, Becquet P, et al. Invited review: Current enteric methane mitigation options. Journal of Dairy Science. 2022;**105**(12):9297-9326. DOI: 10.3168/jds.2022-22091

[124] Benchaar C, Pomar C, Chiquette J. Evaluation of dietary strategies to reduce methane production in ruminants: A modelling approach. Canadian Journal of Animal Science. 2001;**81**(4):563-574. DOI: 10.4141/A00-119

[125] Mosier AR, Duxbury JM, Freney JR, Heinemeyer O, Minami K, Johnson DE. Mitigating agricultural emissions of methane. Climatic Change. 1998;**40**(1):39-80. DOI: 10.1023/A:1005338731269

[126] Haque MN. Dietary manipulation: A sustainable way to mitigate methane emissions from ruminants. Journal of Animal Science and Technology. 2018;**60**(1):15. DOI: 10.1186/s40781-018-0175-7

[127] Kolling GJ, Stivanin SCB, Gabbi AM, Machado FS, Ferreira AL, Campos MM, et al. Performance and methane emissions in dairy cows fed oregano and green tea extracts as feed additives. Journal of Dairy Science. 2018;**101**(5):4221-4234. DOI: 10.3168/jds.2017-13841

[128] Hernández A, Kholif AE, Elghandour MMMY, Camacho LM, Cipriano MM, Salem AZM, et al. Effectiveness of xylanase and *Saccharomyces cerevisiae* as feed additives on gas emissions from agricultural calf farms. Journal of Cleaner Production. 2017;**148**:616-623. DOI: 10.1016/j.jclepro.2017.01.070

[129] Park KY, Jo YH, Ghassemi Nejad J, Lee JC, Lee HG. Evaluation of nutritional value of Ulva sp. and *Sargassum horneri* as potential eco-friendly ruminants feed. Algal Research. 2022;**65**:102706. DOI: 10.1016/j.algal.2022.102706

[130] Wang K, Xiong B, Zhao X. Could propionate formation be used to reduce enteric methane emission in ruminants? Science of the Total Environment. 2023;**855**:158867. DOI: 10.1016/j.scitotenv.2022.158867

[131] Almeida AK, Hegarty RS, Cowie A. Meta-analysis quantifying the potential of dietary additives and rumen modifiers for methane mitigation in ruminant production systems. Animal Nutrition. 2021;**7**(4):1219-1230. DOI: 10.1016/j.aninu.2021.09.005

[132] Santoso B, Mwenya B, Sar C, Gamo Y, Kobayashi T, Morikawa R, et al. Effects of supplementing galacto-oligosaccharides, *Yucca schidigera* or nisin on rumen methanogenesis, nitrogen and energy metabolism in sheep. Livestock Production Science. 2004;**91**(3):209-217. DOI: 10.1016/j.livprodsci.2004.08.004

[133] Min BR, Solaiman S, Waldrip HM, Parker D, Todd RW, Brauer D. Dietary mitigation of enteric methane emissions from ruminants: A review of plant tannin mitigation options. Animal Nutrition. 2020;**6**(3):231-246. DOI: 10.1016/j.aninu.2020.05.002

[134] Waghorn GC, Tavendale MH, Woodfield DR. Methanogenesis from

forages fed to sheep. Proceedings of the New Zealand Grassland Association. 2002;**64**:167-171. DOI: 10.33584/jnzg.2002.64.2462

[135] Woodward SL, Waghorn GC, Ulyatt MJ, Lassey KR. Early indications that feeding lotus will reduce methane emissions from ruminants. New Zealand Society of Animal Production. 2001;**61**:23-26

[136] Bhatta R, Uyeno Y, Tajima K, Takenaka A, Yabumoto Y, Nonaka I, et al. Difference in the nature of tannins on *in vitro* ruminal methane and volatile fatty acid production and on methanogenic archaea and protozoal populations. Journal of Dairy Science. 2009;**92**(11):5512-5522. DOI: 10.3168/jds.2008-1441

[137] Jayanegara A, Goel G, Makkar HPS, Becker K. Divergence between purified hydrolysable and condensed tannin effects on methane emission, rumen fermentation and microbial population *in vitro*. Animal Feed Science and Technology. 2015a;**209**:60-68. DOI: 10.1016/j.anifeedsci.2015.08.002

[138] Anantasook N, Wanapat M, Cherdthong A. Manipulation of ruminal fermentation and methane production by supplementation of rain tree pod meal containing tannins and saponins in growing dairy steers. Journal of Animal Physiology and Animal Nutrition. 2014;**98**(1):50-55. DOI: 10.1111/jpn.12029

[139] Bhatta R, Saravanan M, Baruah L, Sampath KT. Nutrient content, *in vitro* ruminal fermentation characteristics and methane reduction potential of tropical tannin-containing leaves. Journal of the Science of Food and Agriculture. 2012;**92**(15):2929-2935. DOI: 10.1002/jsfa.5703

[140] Bueno ICS, Brandi RA, Franzolin R, Benetel G, Fagundes GM, Abdalla AL, et al. *In vitro* methane production and tolerance to condensed tannins in five ruminant species. Animal Feed Science and Technology. 2015;**205**:1-9. DOI: 10.1016/j.anifeedsci.2015.03.008

[141] Hassanat F, Benchaar C. Assessment of the effect of condensed (acacia and quebracho) and hydrolysable (chestnut and valonea) tannins on rumen fermentation and methane production *in vitro*. Journal of the Science of Food and Agriculture. 2013;**93**(2):332-339. DOI: 10.1002/jsfa.5763

[142] Hatew B, Stringano E, Mueller-Harvey I, Hendriks WH, Carbonero CH, Smith LMJ, et al. Impact of variation in structure of condensed tannins from sainfoin (*Onobrychis viciifolia*) on in vitro ruminal methane production and fermentation characteristics. Journal of Animal Physiology and Animal Nutrition. 2016;**100**(2):348-360. DOI: 10.1111/jpn.12336

[143] Jayanegara A, Kreuzer M, Wina E, Leiber F. Significance of phenolic compounds in tropical forages for the ruminal bypass of polyunsaturated fatty acids and the appearance of biohydrogenation intermediates as examined in vitro. Animal Production Science. 2011;**51**(12):1127. DOI: 10.1071/AN11059

[144] Jayanegara A, Goel G, Makkar HPS, Becker K. Reduction in methane emissions from ruminants by plant secondary metabolites: Effects of polyphenols and saponins. In: Odongo NE, Garcia M, Viljoen GJ, editors. Sustainable Improvement of Animal Production and Health. Animal Production and Health Subprogramme, Joint FAO/IAEA Division of Nuclear Techniques in Food and Agriculture, Department of Nuclear Sciences and Applications, International Atomic

Agency, Vienna, Austria; 2010. pp. 151-157

[145] Naumann HD, Lambert BD, Armstrong SA, Fonseca MA, Tedeschi LO, Muir JP, et al. Effect of replacing alfalfa with panicled-tick clover or sericea lespedeza in corn-alfalfa-based substrates on *in vitro* ruminal methane production. Journal of Dairy Science. 2015;**98**(6):3980-3987. DOI: 10.3168/jds.2014-8836

[146] Pinski B, Günal M, AbuGhazaleh AA. The effects of essential oil and condensed tannin on fermentation and methane production under *in vitro* conditions. Animal Production Science. 2016;**56**(10):1707. DOI: 10.1071/AN15069

[147] Rira M, Chentli A, Boufenera S, Bousseboua H. Effects of plants containing secondary metabolites on ruminal methanogenesis of sheep *in vitro*. Energy Procedia. 2015;**74**:15-24. DOI: 10.1016/j.egypro.2015.07.513

[148] Soltan YA, Morsy AS, Sallam SMA, Lucas RC, Louvandini H, Kreuzer M, et al. Contribution of condensed tannins and mimosine to the methane mitigation caused by feeding *Leucaena leucocephala*. Archives of Animal Nutrition. 2013;**67**(3):169-184. DOI: 10.1080/1745039X.2013.801139

[149] Soltan Y, Morsy A, Sallam S, Louvandini H, Abdalla A. Comparative *in vitro* evaluation of forage legumes (prosopis, acacia, atriplex, and leucaena) on ruminal fermentation and methanogenesis. Journal of Animal and Feed Sciences. 2012;**21**(4):759-772. DOI: 10.22358/jafs/66148/2012

[150] Tan HY, Sieo CC, Abdullah N, Liang JB, Huang XD, Ho YW. Effects of condensed tannins from Leucaena on methane production, rumen fermentation and populations of methanogens and protozoa *in vitro*. Animal Feed Science and Technology. 2011;**169**(3-4):185-193. DOI: 10.1016/j.anifeedsci.2011.07.004

[151] Wanapat M, Chanthakhoun V, Phesatcha K, Kang S. Influence of mangosteen peel powder as a source of plant secondary compounds on rumen microorganisms, volatile fatty acids, methane and microbial protein synthesis in swamp buffaloes. Livestock Science. 2014;**162**:126-133. DOI: 10.1016/j.livsci.2014.01.025

[152] Wischer G, Boguhn J, Steingaß H, Schollenberger M, Rodehutscord M. Effects of different tannin-rich extracts and rapeseed tannin monomers on methane formation and microbial protein synthesis *in vitro*. Animal. 2013;7(11):1796-1805. DOI: 10.1017/S1751731113001481

[153] Klita PT, Mathison GW, Fenton TW, Hardin RT. Effects of alfalfa root saponins on digestive function in sheep. Journal of Animal Science. 1996;**74**(5):1144. DOI: 10.2527/1996.7451144x

[154] Lu CD, Jorgensen NA. Alfalfa saponins affect site and extent of nutrient digestion in ruminants. The Journal of Nutrition. 1987;**117**(5):919-927. DOI: 10.1093/jn/117.5.919

[155] Jouany J-P. Effect of rumen protozoa on nitrogen utilization by ruminants. The Journal of Nutrition. 1996;**126**:1335S-1346S. DOI: 10.1093/jn/126.suppl_4.1335S

[156] Jouany J-P, Morgavi DP. Use of 'natural' products as alternatives to antibiotic feed additives in ruminant production. Animal. 2007;**1**(10):1443-1466. DOI: 10.1017/S1751731107000742

[157] Bharathidhasan A, Viswanathan K, Balakrishnan V, Valli C, Ramesh S, Senthilkumar SMA. Effects of purified saponin on rumen methanogenesis and rumen fermentation characteristics studied using *in vitro* gas production technique. International Journal of Veterinary Science. 2013;**2**(2):44-49

[158] Jafari S, Goh YM, Rajion MA, Faseleh Jahromi M, Ebrahimi M. Ruminal methanogenesis and biohydrogenation reduction potential of papaya (Carica papaya) leaf: An *in vitro* study. Italian Journal of Animal Science. 2016a;**15**(1):157-165. DOI: 10.1080/1828051X.2016.1141031

[159] Jafari S, Meng GY, Rajion MA, Jahromi MF, Ebrahimi M. Manipulation of rumen microbial fermentation by polyphenol rich solvent fractions from papaya leaf to reduce green-house gas methane and biohydrogenation of C18 PUFA. Journal of Agricultural and Food Chemistry. 2016b;**64**(22):4522-4530. DOI: 10.1021/acs.jafc.6b00846

[160] Jafari S, Goh YM, Rajion MA, Jahromi MF, Ahmad YH, Ebrahimi M. Papaya (*Carica papaya*) leaf methanolic extract modulates *in vitro* rumen methanogenesis and rumen biohydrogenation. Animal Science Journal. 2017;**88**(2):267-276. DOI: 10.1111/asj.12634

[161] Li W, Powers W. Effects of saponin extracts on air emissions from steers 1. Journal of Animal Science. 2012;**90**(11):4001-4013. DOI: 10.2527/jas.2011-4888

[162] Narvaez N, Wang Y, McAllister T. Effects of extracts of Humulus lupulus (hops) and *Yucca schidigera* applied alone or in combination with monensin on rumen fermentation and microbial populations in vitro. Journal of the Science of Food and Agriculture. 2013;**93**(10):2517-2522. DOI: 10.1002/jsfa.6068

[163] Patra AK, Yu Z. Effective reduction of enteric methane production by a combination of nitrate and saponin without adverse effect on feed degradability, fermentation, or bacterial and archaeal communities of the rumen. Bioresource Technology. 2013;**148**:352-360. DOI: 10.1016/j.biortech.2013.08.140

[164] Patra AK, Yu Z. Combinations of nitrate, saponin, and sulfate additively reduce methane production by rumen cultures in vitro while not adversely affecting feed digestion, fermentation or microbial communities. Bioresource Technology. 2014;**155**:129-135. DOI: 10.1016/j.biortech.2013.12.099

[165] Patra AK, Yu Z. Effects of adaptation of In vitro rumen culture to garlic oil, nitrate, and saponin and their combinations on methanogenesis, fermentation, and abundances and diversity of microbial populations. Frontiers in Microbiology. 2015;**6**:1434. DOI: 10.3389/fmicb.2015.01434